CARNIVOROUS PLANTS

Reaktion's Botanical series is the first of its kind, integrating horticultural and botanical writing with a broader account of the cultural and social impact of trees, plants and flowers.

CARNIVOROUS PLANTS

Dan Torre

REAKTION BOOKS

Published by
REAKTION BOOKS LTD
Unit 32, Waterside
44–48 Wharf Road
London N1 7UX, UK

www.reaktionbooks.co.uk

First published 2019
Copyright © Dan Torre 2019

Printed and bound in China by 1010 Printing International Ltd

A catalogue record for this book is available from the British Library

ISBN 978 1 78914 052 1

Contents

Sarracenia or North American pitcher plant.

Introduction

❦

Carnivorous plants represent one of the most extraordinary groupings within the plant kingdom. They comprise a wide variety of plants that have evolved specialized leaves to attract, trap, kill and consume insects, spiders and other small creatures. They have attracted and captured our imaginations, providing inspiration for a wide range of art, literature, cinema, animation and popular culture.

The very concept of a carnivorous plant seems at first to be incompatible with our traditional view of the natural world. We expect plants to be at the very base of the food chain, providing nourishment for animals – not the other way round. Certainly this was a puzzling feature throughout the eighteenth and nineteenth centuries when it was first suggested that these plants might be carnivores. Such a preposterous claim seemed directly to contradict the accepted fundamental ordering of the natural world. Even today, the idea that plants can 'eat meat' seems rather strange to many of us – which helps to explain why there are so many quirky references to these plants in popular culture.

Over 700 species of plants are currently recognized as being carnivorous. And their numbers keep increasing – each year new species are discovered and old ones are identified as being carnivorous. Interestingly, the definition of what makes a plant carnivorous has gradually been amended over the decades, and this also has caused their numbers to grow. It is generally agreed that a carnivorous plant

Illustration of assorted carnivorous plants from *Meyers Konversations-Lexikon* (1897 edn).

must have the capacity to attract, trap, kill and digest insects, and then to be able to derive nutritional benefit from them. Nevertheless, many carnivorous plants will rely upon bacteria or even other insects to help with the actual digestion.

Some of these species have been known for many decades, even centuries, but only more recently determined to be carnivorous. Some plants were well known to have the ability to catch insects, but this was often thought to be either accidental or purely for defence. For example, the pitcher plant was thought to be able to catch rain in order to store up a reserve of water, with unfortunate insects sometimes drowning while trying to steal a drink. Others, such as the Venus flytrap or the sundew, were thought to catch insects as a defensive mechanism – if the plant killed the insect first, then it would not be able to munch on its leaves and flowers. However, we now recognize that eating insects is of primary significance to these plants. Significantly, a survey of carnivorous plants reveals that there are rarely any predetermined boundaries in the natural world. Many plants might exhibit some carnivorous traits, yet are not recognized as being fully carnivorous. For example, they might capture and kill insects, but not actively consume and gain direct nutritional benefit from them. Such plants are often referred to as semi-carnivorous.

Plants, of course, absorb water from the ground; but they also absorb a wide range of nutrients (particularly potassium and nitrogen) through their roots, and quite often through their leaves. Frequently, these essential nutrients will have been derived directly or indirectly from other decaying plants and animals. In fact, many natural fertilizers are derived either from animal manure or from dead animals (as in 'blood and bone' fertilizers). So virtually all plants benefit from either the carcasses or excrement of animals – but are not, of course, considered to be carnivorous. Similarly, most plants can benefit from the rotting remains of other plants, absorbing these nutrients through their roots, but we rarely think of them as being what they truly are – cannibals.

Darlingtonia californica, or cobra plant, with its characteristic translucent domed pitcher and protruding 'tongue'.

Although most carnivorous plants can survive and grow reasonably well without any 'meat' at all, they will usually exhibit significant growth when it is available. This is, of course, variable, and some carnivorous plants will rely more on captured prey than others. Some plants, such as *Triphyophyllum* (an enormous vine that can grow over 45 m (148 ft) in length), are carnivorous for only a brief time during their regular life cycle.

Much like current human trends, some carnivorous plants seem to be going vegetarian, eschewing meat and relying instead on falling leaf matter and airborne pollen from other plants. A few species of pitcher plants (*Nepenthes*) have even taken to consuming primarily bat and rat excrement in their pitchers for nutritional gain.

Carnivorous plants can be found in nearly every corner of the Earth, from the swamps of the eastern United States to the chilliest British and European highlands, the dampest tropical jungles of Southeast Asia and even, surprisingly, the arid lands of Mexico and Australia. But what nearly all these environments have in common are growing conditions lacking in essential nutrients. The plants compensate by capturing and eating insects, thereby gaining a distinct advantage over other plants that are restricted to absorbing nutrients from the soil. Remarkably, carnivorous plants do not necessarily kill every type of available insect. In fact, many varieties have formed incredibly complex relationships with certain insects and animals.

Carnivorous plants are a remarkable vegetation that have a big reputation – one that far surpasses their normally diminutive size. This larger-than-life reputation is often humorously expressed in both their creative depiction and in their botanical display. Recently, at the Singapore Botanic – Gardens at the Bay, an amusing display of carnivorous plants was installed. It appeared, at first glance, to feature gargantuan pitcher plants and Venus flytraps – so large as to be capable of swallowing a small dog or cat. On closer inspection, it became obvious that these were merely plastic sculptures made from

Sarracenia pitcher plant cultivars.

Carnivorous plant display (along with numerous Lego-block sculptures) in Singapore.

Lego-style blocks that had been playfully intermingled with the real-life, and normal-sized, carnivorous plants. Although the display was obviously facetious and exaggerated, it also managed to highlight what many might hope to see in a carnivorous plant display – big man-eating plants!

Since their initial discovery, carnivorous plants have inspired our imaginations. Quite often we have thought of them as killers, and countless films, animations, comics and books have depicted them in this manner. However, there have also been many subtler and more complimentary representations, ranging from depicting them as cute and intelligent pets, to exotic and simply beautiful forms. Recently there has been an accelerating interest in the growing, collecting and conserving of these remarkable plants. Ultimately, carnivorous plants occupy a unique space within the plant kingdom and an equally unique space within our imaginations.

one

The Natural History of Carnivorous Plants

❦

C arnivorous plants comprise some of the most unusual species on the planet. There are currently about 700 known
species, and new varieties are constantly being found – on
average some ten species per year have been discovered over the
last decade. But as there are, in total, over 390,000 species of vas-
cular plants, carnivority remains a specialized characteristic within
the plant kingdom. These carnivorous tendencies constitute a unique
but informal taxonomic grouping. Currently, carnivorous plants are
known to span nearly twenty different genera and about eleven
different families. They are believed to have derived from at least
six different lineages.[1]

The invariable characteristic these plants have in common is that
they live in nutrient-poor growing conditions. Thus instead of devoting
their energy to growing large root systems in an attempt to compete
with other plants for an extremely limited supply of nutrients, they
have instead developed modified leaves that are capable of trapping,
killing and then extracting these much-needed nutrients from insects
and other small animal creatures. There are numerous ways in which
they go about being carnivorous, and over the decades there has been
debate as to exactly what classifies a plant as being truly so. Generally,
it is agreed that the plant will need to attract and trap insects, digest
the insects (with or without assistance from other life-forms) and
absorb and benefit from the nutrients of those digested insects. Over
the decades, it has usually been the eating/digesting that has been the

most contentious aspect of the definition of the carnivorous plant grouping. In the past, some have insisted that the plant must be capable of digesting its prey entirely on its own – such as the Venus flytrap. Others have asserted that plants like the sun pitcher (*Heliamphora*) or the cobra plant (*Darlingtonia*), both of which rely almost

Illustration of various carnivorous plants from *Brockhaus' Konversations-Lexikon* (1892).

exclusively on bacteria and other microorganisms to break down their prey before they absorb it, are equally as carnivorous as the flytrap. It is now mostly agreed that it is perfectly acceptable for plants to get some help in this area and still be considered carnivorous. By comparison, animals, including humans, rely upon bacteria living within their intestines to properly digest their food – which doesn't discredit them from being considered 'capable of digestion'.

It is undeniable that virtually all plants will benefit nutritionally if a dead animal decomposes at the base of their roots. But while there are many plants that might have some of the characteristics of the carnivorous plants, what makes carnivorous plants special is that they, to put it in very human terms, appear to have carnivorous intent. They seem to *want* to eat insects and they will often go to great lengths to lure, catch, digest and then absorb them. Again, while this may be a somewhat anthropocentric way to consider it, it is also an effective way to distinguish and categorize a very diverse, but very special, group of plants.

There are generally good reasons why these plants seem to buck the trend of most others and resort to being carnivorous, which generally have to do with the rather inhospitable environments in which they find themselves. Carnivority seems to be the efficient way to overcome the hardships of their surroundings and give them a competitive advantage over other plants in the locality. One writer, speaking in almost purely economic terms, argues that carnivorousness requires a great deal of effort on behalf of the plant; therefore, only those in the poorest of soils and with ample prey will find it worth their while:

> Cost-effectiveness is the key determinant. The substantial investments a plant must make to produce the traps and other accoutrements required to capture and process prey must be paid back, and profits gained after that. This is not possible where roots, which are cheaper, have access to abundant supplies of key nutrients like nitrogen and phosphorus.[2]

All carnivorous plants primarily use their leaves to assist with their carnivorous nature. Even the remarkable corkscrew plant (*Genlisea*), which eats underground insects, uses its leaves (which look like roots) to carry out the necessary trapping and digesting. It is this shift from feeding via root to feeding via leaves that stands out as one of the most intriguing aspects of carnivorous plants.

It is, of course, erroneous to think that non-carnivorous plants are only able to absorb nutrients through their roots. Many are capable of foliar feeding, of being able to assimilate nutrients through their leaves. Fertilizer sprayed onto the leaves of a plant, for example, will most probably be quickly absorbed into the plant and utilized.[3] But carnivorous plants developed an ultra-efficient means of using their leaves in this way. It is believed that, over time, some leaves developed curvatures and indentations that allowed rainfall to collect and in which occasional insects would drown. To the leaves, already able to absorb nutrients, these decomposing insects would serve as an extra foliar fertilizer.[4] Consequently, plant leaves with deeper 'cups' would be able to catch more insects and thus gain more foliar nutrients.[5]

It has also been suggested that even when there are healthy amounts of nitrogen and other nutrients in the soil, many carnivores, such as the pitcher plants *Sarracenia pupurea* and *Darlingtonia*, have become very inefficient in absorbing nutrients through their roots. It is likely that many species have become more adept at foliar absorption at the expense of root-absorption capabilities.[6] *Sarracenia* absorb as much as 80 per cent of their nitrogen requirements directly into their pitcher leaves. Even when seasonal rainfall is highly rich in nitrogen, *Sarracenia pupurea* will primarily absorb its nutrients from the rainwater that has collected inside its pitchers rather than absorbing this through its roots like most other plants.[7] Some carnivorous plants, such as the waterwheel plant (*Aldrovanda*), the corkscrew plant (*Genlisea*) and bladderworts (*Utricularia*), have no roots at all; the last two have leaves that, because they normally grow underground, have taken on the aesthetics of roots.

Detail of a *Drosophyllum* with captured prey.

It is believed that Venus flytraps (*Dionaea*) shared a common ancestor with sundews (*Drosera*) and subsequently developed more advanced trapping mechanisms. Some have suggested that a motivation for this development could be the ability to capture larger prey, and thus gain a greater reward.[8] A snap-trap can more effectively catch a larger insect than can a sticky leaf. Some *Drosera*, however, have developed snap-tentacles, which might be considered precursors to the development of *Dionaea*-like snap-traps.

Most carnivorous plants hail from open spaces where there is naturally full sun with little shade. They tend to grow in very nutrient-poor soils in which larger shade-producing plants would be unable to survive.[9] In addition to poor soil and direct sunlight, many, such as *Sarracenia*, *Dionaea*, *Darlingtonia* and some *Drosera*, also come from temperate climates where they have been accustomed to a dormancy period. During the cold winter months, many of these plants will die back to their roots, surviving in a bulb-like form.[10] Some carnivorous plants, especially North American varieties, are well equipped to withstand periodic fires and have, in fact, come to rely upon fire in order to flourish. Because they generally grow in open spaces and are not very good at competing with other plants for space and light, a bush fire will effectively clear away their competitors while they bide time underground. Then, when the smoke has cleared, they grow again and flourish.

There are many different varieties of carnivorous plants, each employing a wide range of strategies for capturing their prey. Most can be encompassed within a few thematic groupings based on their preferred method of capture. This chapter is therefore organized into three key sections: fast-moving carnivores, sticky carnivores and pitcher plant carnivores.

Fast-moving Carnivores

Most plant life will grow or move at a rate that is invisible to the human eye. However, there are a number of carnivorous plant species that can exhibit sudden snap-trap movements – these are not only highly visible, but could be described as fast-moving, even by human standards.

Venus Flytrap (*Dionaea muscipula*)

The Venus flytrap is unquestionably the most famous of all the carnivorous plants. It is also the most theatrical in the way that it swiftly captures its prey with its spectacular snap-trap leaves. Contrary to what its name suggests, its diet primarily consists of crawling creatures, such as ants and spiders – even the occasional froglet or small lizard. Studies have shown that flying insects make up less than 25 per cent of its natural habitat diet.[11] *Dionaea muscipula* is the sole species in the genus *Dionaea*, which is part of the family Droseraceae. This larger family category also includes the sundew, and the two are often seen growing alongside each other in habitat.

Venus flytraps are quite small in stature; they will generally form a rosette of low-lying leaves, roughly 10 cm (4 in.) in diameter, which will sprout from an underground bulb-like rhizome. Its leaves are modified at the ends, being split into two lobes that form an open 'clam shell' with spiky 'teeth' that make up its highly effective trigger-traps. It is able to attract a wide range of insects, and does so through various means including bright coloration, special UV patterning and fragrance – the traps produce an alluring essence that is secreted from glands along the outer edges of the open traps, to which insects are attracted.

The interior of each trap has six small protruding hairs – three on each half-surface – and it is these that stimulate the snapping action, enabling the traps to snap shut in a fraction of a second. And to avoid unnecessary and costly closures, the plant has developed an ingenious means by which to ensure that it has not captured a falling leaf or other such non-edible matter. To close, the trap normally requires two

Overleaf: Venus flytrap cultivar with vibrant red-coloured snap-traps.

Venus flytrap (*Dionaea muscipula*) cultivar with upright traps
visibly expressing different stages of its hunting cycle.

hairs to have been triggered within about a twenty- to 25-second
time frame, or, alternatively, for a single hair to be triggered twice
within this same time frame. The bending of a trigger-hair sends an
action potential (an electrical impulse) to the motor cells at the base
of the trap, where it is stored; then, when a second impulse is received
(within twenty seconds or so), it activates the trap to shut. If no
second impulse is received, then it 'forgets' and releases the stored
energy. After which, it will again wait for a new set of impulses. Such
a process can best be described as a form of 'short term electrical
memory'.[12] Although many plants are capable of much longer-term
memory, limiting this trapping process to a short-term memory span
of about twenty seconds is ideal. If its memory were any longer then
it would, for example, remember each accidental trigger (say from a
falling drop of water) and would be continually snapping shut its traps,
wasting enormous amounts of energy in the process.[13] Researchers
have also found that pressing down on a single hair for a sustained
length of time (more than two seconds) will also trigger the closure
of the traps as multiple charges (action potentials) will be sent to the

motor cells.[14] Furthermore, proving that they act on electrical impulses, the traps can also be made to close by introducing electrical stimuli near the midrib of the trap.[15]

The prevailing theory as to how the traps physically close can be described by what is known as the 'hydroelastic curvature model'. In response to the trigger, a flow of water occurs between neighbouring cells which 'quickly changes [the leaf's] curvature from convex to concave and the trap closes'.[16] Because of the concave shape of the newly snapped trap, the flytrap does not initially crush its prey, but merely imprisons it – leaving small gaps between the teeth. Within the next five minutes, while the plant 'decides' whether or not the prey is suitable, a small amount of further tightening or closure occurs. This decision-making process is largely based upon electrical signals that continue to be produced by the stimulation of the trigger hairs by the struggling insect. Once the plant has decided that the prey is worthy of consumption, it will commence digestion by sealing shut the traps and subtly flattening down the concave form of the lobes.[17] In this newly watertight chamber it will then begin to exude digestion juices, which start to dissolve and absorb the soft tissue and fluids of the insect. Digestion normally takes between five and ten days but can take longer in some instances. Once digestion is complete, the trap will slowly open over a period of 24 hours, making visible the remaining undigested hull of the insect. Initially, the lobes will still be in a concave shape; another day is required before the lobes will bend outward into the fully open convex shape. Normally during this period, the husk of the insect will either be blown away in the wind or flushed away in the rain. However, the trigger hairs will remain desensitized for a couple of days more, so that any remaining carcass will not trigger the trap closure. Finally, the trigger hairs of the trap will become sensitive again.

In their natural habitat (which is very small and localized in only one area of the southeastern United States) the plants normally grow in wet, swampy areas. If too dry, the plants will die back to the soil, but their underground rhizome bulbs will persist. In flooded conditions, the plants can usually survive and feed – even when completely submerged in water for some time. In fact, some flytraps have even been observed actively catching underwater prey, such as flatworms, when the traps are fully submerged.[18] The plants are very winter hardy and will normally go into dormancy when the local temperatures drop well below freezing.[19] They produce white flowers that bloom in early summer. The flower stalks stretch up, far above the low-lying traps, growing up to 40 cm (16 in.) tall. This tall growth of the inflorescence is very important, ensuring that the plant does not accidentally eat one of its pollinators.

Like all plants, Venus flytraps are susceptible to leaf damage by hungry herbivores. Most plants, if an insect eats a portion of one of their leaves, are likely to remain perfectly viable, still able to continue their essential role of providing food through photosynthesis and

Venus flytraps in bloom.

foliar absorption of nutrients. But a hole or other such damage to the Venus flytrap would immediately render the leaf-trap useless. So the flytrap needs to be very vigilant in discouraging predators, especially since it is so expensive to create its specialized leaf-traps. In order to counter this threat, Venus flytraps, like many other plants, are able to produce slightly toxic chemicals in their leaves that can discourage herbivore insects from eating them. But because most of the volatile chemicals that other plants produce are nitrogen-based, this presents a problem for the Venus flytrap, which grows in very nitrogen-starved environments. As a result it has developed ingenious non-nitrogen based chemicals (such as phenolics and flavonol glycosides) to discourage its leaf predators. Furthermore, the traps also produce a slightly volatile chemical known as plumbagin, which is believed to have a narcotic effect on the trapped insects – sedating them so that, before they are killed by suffocation in the digestive fluids, their continuing struggling will not damage the precious leaf-traps.[20]

The Waterwheel Plant (*Aldrovanda vesiculosa*)

Aldrovanda vesiculosa is another fast-moving carnivorous plant, commonly known as the waterwheel plant owing to its leaves being arranged in a whorl formation of six to nine leaves which radiate like the spokes of a wheel. There are usually a number of these whorls situated along the main stem of the plant. *Aldrovanda* is a rootless, free-floating, freshwater plant that normally remains submerged just below the surface.

On the tip of each leaf sits a snap-trap very similar in design to the Venus flytrap but much smaller, averaging only about 2 mm in length. The walls of the traps are very thin and mostly translucent. Whereas the Venus flytrap has only six trigger hairs per trap, the waterwheel plant has approximately twenty to forty (or more) inside its traps. Because of the plant's broad distribution, its diet can vary widely, but normally it consists of fly larvae and water fleas and also eelworms, which will swim into the traps and trigger their closure.[21] Once the trap closes with its prey inside, digestive juices are secreted from sessile

Water wheel plant (*Aldrovanda vesiculosa*), which grows partly submerged on the surface of lakes and ponds.

glands on the inside surfaces of the trap. Its hunting cycle is very similar to that of the Venus flytrap. The traps will initially snap shut very quickly, and then slowly constrict (expelling any trapped water) and seal its edges. It will then secrete digestive juices and slowly digest and absorb the insect. After several days it will open its trap, allowing any insect remains to drift away before it becomes active again. As with the Venus flytrap, each trap can be activated several times before it dies.

A mature plant can reach a maximum length of about 30 cm (12 in.). Although it grows continuously, it maintains this optimum length by concentrating its growth on just one end; as it ages the other end withers away, thereby ensuring continual new growth but of a fairly consistent and sustainable length. *Aldrovanda* tends to live in still or slowly flowing water and is propagated primarily from pieces that get broken off the main plant. This often occurs when waterfowl swim through the plant; they will invariably carry the broken pieces some distance away before they become dislodged.

Bladderwort (*Utricularia*)

A third type of fast-moving carnivore is the bladderwort, which belongs to the genus *Utricularia*. The name 'Utricularia' comes from the Latin *utriculus*, meaning a 'leather-skin bag'; the plant's bladders are somewhat reminiscent of traditional wine-skins. There are more than 230 known species of bladderworts that are distributed widely across the world on every continent except Antarctica, but with the highest concentration in South America. They are the largest genera and also the fastest-moving of carnivorous plants. They are all rootless, having modified leaves (often looking like roots) with small bladder sacks attached. It is with these bladders that they capture their prey. The genus is divided into three main categories: terrestrial (land dwelling), aquatic (water dwelling) and epiphytic (tree-branch dwelling) plants.

The bladder traps, hollow oval forms with an entrance 'door' at one end, are very small, ranging from about 1 mm to 1 cm (0.04 to 0.4 in.) in diameter, and are used to capture equally small creatures. When sealed shut, the bladder creates a pressurized vacuum. Bladderworts

Tiny bladder traps of *Utricularia aurea*.

also have a trigger hair that, when touched by an insect, causes the trap door to open inwards, sucking in water along with the insect into the empty bladder. Once the insect is inside, the door slams shut again. This trapping process happens very quickly, at about two thousandths of a second, making it one of the fastest-moving plants in the world. The bladder then expels the trapped water, and the

Illustration of a bladderwort (*Utricularia*) growing in water, *c.* 1912.

walls of the bladder compress around the trapped insect, which it then digests.

Wherever bladderworts grow, they require that their bladder-traps be submerged in water in order for them to function. The terrestrial species (which make up the bulk of the genus) are therefore usually located in soils that remain waterlogged for most of the growing season.[22] The aquatic species are generally found free-floating in water. The epiphytic species normally grow upon water-soaked mossy patches on the branches of trees in very wet tropical areas. There is even one species, *Utricularia humboldtii*, which grows within the cistern-like leaves of the bromeliad plant. These can grow quite large, sending up flower stems over a metre tall.[23] Many bladderworts have impressive flowers which can be as large as 5 cm (2 in.) in diameter.

Sticky Carnivores

Many carnivorous plant species exude a glue-like substance on the surface of their leaves, which helps to attract and capture insects. The unfortunate insects that get trapped in this adhesive fluid will almost certainly meet a very sticky end.

Sundew (*Drosera*)

Drosera, or sundews, are a very widespread genus of carnivorous plant. There are nearly two hundred different species inhabiting nearly every continent on earth. Approximately half of the species are native to Australia. While most species grow in wet, low-nutrient soils, some have adapted to areas that can become very dry during the summer months. These varieties generally produce underground bulbous structures that allow them to die back to the soil level and thus survive underground during extremely dry periods, and even through bush fires.

Although their leaves can come in just about any imaginable shape and size, from flat and round, to long, thin and spindly, all are covered

Drosera rotundifolia, forming small compact rosettes.

in tiny hair-like tentacles that have a small gland at the tip. From this gland, a very sticky droplet of mucilage is produced, which forms into what looks like a dewdrop. Appropriately, the name 'Drosera' comes from the Greek term *droseros* meaning 'dewy'. However, unlike regular dew from condensed water, which will quickly evaporate in the sun, the hotter and sunnier it becomes the greater the quantities of 'dew' the sundews seem to produce.

These droplets of dew-like mucilage are very attractive to insects as they reflect both visible and ultraviolet light spectrums to resemble a sweet nectar. The insect will alight on the leaves in an attempt to feed on them. But because the mucilage is so sticky, any insect that touches it with its legs, wings or body will immediately become stuck, unable to free itself. As it tries to wrestle free, it will usually become even more firmly stuck as its thrashing body comes into contact with more of the sticky tentacles.

Many sundews are not just passive flypaper traps but are quite reactive in their capturing of insects in that, as soon as one becomes caught in its glue, nearby tentacles will begin to bend slowly towards the insect. Some species, especially *D. glanduligera*, possess very fast-moving 'snap-tentacles' along the edges of their leaves. These will instantly (within a tenth of a second) bend over and press the insect into the sticky centre of the leaf. In some other species the entire leaf will, over a period of several hours, slowly curl around the insect. The fast movements of the snap-tentacles are, like the Venus flytrap, the result of electrical impulses, while the slower movements are likely to be a result of chemical stimulation, caused probably by an increased production of hormones.[24] Soon after the insect is caught within the sundew's sticky leaves, the tentacles will stop producing mucilage and instead begin to secrete acidic fluids comprising digestive enzymes. Once parts of the insect have been digested, the same glands will then begin to absorb the now liquefied insect matter. After a few days, all that will be left is an empty exoskeleton. Because the nearby

Sundew (*Drosera binata*), with its many tiny hair-like tentacles, each with a sticky 'dewdrop' of mucilage.

tentacles stop producing mucilage after digestion begins, the remains of the insect will normally fall away when the plant has finished with it. Only then will the tentacles resume production of their sticky glue.

Sundews can range greatly in size. The smallest might have leaves only a few millimetres long (the entire plant of the *Drosera occidentalis* from Australia is usually only 8 to 10 mm, or a third of an inch, in diameter), while the *Drosera regia* has long thin leaves that can reach more than 50 cm (1½ ft) in length. Some sundews form small rosettes, not unlike the structure of a Venus flytrap, while others are essentially long climbing plants. The flowers range in size and come in a variety of colours: white, red, yellow, orange and purple.

Butterwort (*Pinguicula*)

There are approximately eighty different species of *Pinguicula,* or butterwort. Most can be found in Mexico and Central America but they are also found in North America, Europe and Asia. Like the sundew, butterworts capture their prey with an array of sticky hairs. These hairs tend to be much smaller, and from a distance the leaves can appear to be quite smooth and to have an oily sheen. It is this oily appearance that has earned it the common name of butterwort; similarly, the Latin name *Pinguicula* also refers to oily fat.

Most butterworts are small rosette-forming plants, some 5 to 10 cm (2 to 4 in.) in diameter. The tiny glands that cover the leaves produce the sticky mucilage. Unlike most sundews, butterworts also have leaf surface glands that are situated between the sticky hairs. These surface (or sessile) glands will secrete large amounts of digestive fluids soon after an insect is caught. These same glands will then absorb the insect's nutrients once it has been digested.

When an insect is caught, many butterworts will very slowly begin to curl up the edges of their leaves to form a dish-like area around the insect. Here the digestive fluids will form into little pools, which will ensure better digestion and help to prevent the prey (and the digestive juices) from flowing away. Conveniently, the plants'

The minute sticky insect-catching tentacles that coat the leaves of a butterwort
(*Pinguicula*) give the plants an oily or buttery sheen.

digestive enzymes contain antibacterial qualities that help to keep
the insects fresh while they are being slowly digested.

The Rainbow Plant (*Byblis*)

The rainbow plant genus (*Byblis*), native to Australia and New Guinea,
comprises seven different species. The plants are referred to as
rainbow plants owing to the way that the light hits and colourfully
illuminates the thousands of sticky 'dewdrops' that cover its long
thin leaves.

Like the sundew, these sticky dewdrops of mucilage are produced
by scores of glandular hairs that cover the leaves and stems. *Byblis*
plants have two types of glands; the hairs produce the sticky glue, while
the leaf and stem surface glands (sessile glands) produce the digestive fluids. When an insect is trapped, the sessile glands will produce
the digestive juices, and will then absorb the digested nutrients. These
plants also benefit from helper bugs, *Setocoris bybliphilus* (Miridae),

Byblis filifolia or rainbow plant, native to Australia and New Guinea.

which will eat the insides of some of the larger trapped insects and then excrete them onto the stems and leaves for the plant more readily to absorb.

The Dewy Pine (*Drosophyllum lusitanicum*)

The dewy pine is native only to the western side of the Iberian Peninsula of Spain and Portugal and to the northern areas of Morocco. Its largest concentration is in Portugal and, for this reason, the plant is also commonly referred to as the Portuguese dewy pine. Its scientific name, *Drosophyllum*, 'dewy leaf', comes from the Greek. The dewy pine was first discovered in the seventeenth century and in 1753 Linnaeus initially grouped it with the sundew genus (*Drosera*). Darwin experimented with this species in the 1870s, soon adding it to his long list of insectivorous plants.

The plant normally lives in rather dry soil and grows in areas where the summers can become very hot and dry. It is believed

Dewy pine (*Drosophyllum*), native to the Iberian Peninsula.

that the plants gain much of their water needs from the night-time fog common in the area. They will collect droplets of fog on their mucilage-covered narrow leaves, where it will be readily absorbed.[25] The plants grow narrow stem-like leaves about 15 to 20 cm (6 to 8 in.) tall, somewhat reminiscent of pine needles, particularly when they dry out (thus the common name, dewy pine). A fully mature adult plant can grow as tall as 90 cm (35 in.).

Like the sundew, the dewy pine is covered in scores of glandular hairs, each with a minute drop of sticky mucilage that is tinted a bright red colour. Along the surface of the needle-like leaves are sessile glands that both secrete digestive juices and are capable of absorbing nutrients. Insects are attracted by the glistening red 'dew' and to the plant's alluring honey-like scent.

Once an insect is trapped, it will struggle, becoming covered in sticky glue. The surface glands produce digestive juices which quickly turn the insect into a nutritious soup. This liquid will trickle down the needles, to be absorbed as it flows over a large number of absorption glands.[26] Both the secretion of the digestive glands and the mucilage on the glandular hairs are highly acidic, fully digesting small insects, such as mosquitoes, in as little as 24 hours.[27]

Other Sticky Carnivores

Philcoxia is a very unusual genus of carnivorous plants, which was only recently discovered and even more recently determined, in 2012, to be unequivocally carnivorous. The genus is composed of four species: *P. bahiensis*, *P. goiasensis*, *P. minensis* and the newly discovered *P. tuberosa*. This strange plant keeps its carnivory hidden underground. Above the surface, the plant produces what appears to be a very delicate, spindly and leafless stem. However, this above-ground growth is merely the plant's inflorescence (stalk and very small flowers). Just below the surface grow tiny leaves, just 2 mm (0.08 in.) in diameter, which are covered in sticky hairs. These minuscule leaves capture near microscopic nematode worms that swim through the wet sandy soil.

Once the worm is trapped, the leaves secrete digestive enzymes and the digested worms are then absorbed. The tiny leaves are also able to carry out photosynthesis, because the crystalline, sandy soil that naturally covers the plant allows light to filter down to the leaf surface.

Two other notable varieties of sticky carnivores are *Triphyophyllum peltatum*, a species native to West Africa that produces sticky glandular haired leaves for only a brief portion of its life cycle, making it a uniquely part-time carnivore, and *Roridula*, another sticky-leaved genus which is composed of two species: *R. dentate* and *R. gorgonias*. These plants rely upon host insects (which are immune to the plants' sticky glands) to help digest the less fortunate insects that become captured in the glandular hairs.

Pitcher Plant Carnivores

There is a wide variety of carnivorous pitcher plants — some with very different appearances. But what they all have in common is that they possess fluid-holding receptacles (or pitchers) into which insects are enticed and then drown, subsequently becoming plant-food.

Tropical Pitcher Plants (*Nepenthes*)

Nepenthes is a genus that includes over 150 species and more than 230 recognized hybrids. They are native primarily to Southeast Asia and southern China, but also to northern Australia, Madagascar, southern India and the Seychelles. They are commonly referred to as tropical pitcher plants, hanging pitcher plants or Asian pitcher plants. In Malaysia, they are called *periok kera*, meaning monkey-pot plant, while in China they are commonly referred to as *zhu long cao*, meaning pig-cage plant.[28]

Most *Nepenthes* grow as climbing vines which produce remarkable fluid-filled pitchers. The pitchers are modified leaves that grow in a multi-stage sequence. The plant will initially grow a conventional leaf form, but from the tip of this a long tendril will extend. As it

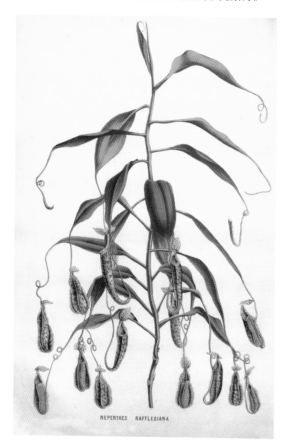

Illustration depicting both upper and lower pitchers of *Nepenthes rafflesiana*, from *Flore des serres et des jardins de l'Europe* (1865).

NEPENTHES RAFFLESIANA

grows longer, the tip of this tendril will begin to enlarge and swell to form a pitcher. When mature, plants will also produce two very distinctive types of pitchers – referred to as lower and upper pitchers. The lower pitchers will grow near the base of the plant – sometimes hanging, sometimes resting upon the ground; they are normally squatter in stature and will often attract crawling insects. The largest pitchers can hold over 3 litres (5.3 pints) of water. Upper pitchers, as the name describes, grow further up the vine and tend to be smaller and more elongated. Because of their higher elevation, the upper pitchers tend to attract a greater number of flying insects. The stems of these pitchers are prehensile and will coil around tree branches for climbing support.[29] The direction in which the pitchers face will normally indicate their type: lower pitchers face towards their supporting leaf and leaf tendril; upper pitchers face away from it. In the taller vines the plants will sometimes form a third kind known as intermediate pitchers, which tend to have a formation somewhere inbetween the upper and lower pitchers. Although there is little difference between the conventional leaves of each species, the pitchers can exhibit remarkable differences.

The *Nepenthes* genus is divided between two groupings based on their natural habitat: lowland species grow at elevations below 1,200 metres (0.75 mi.), while highland species grow above this height. The lowland species generally enjoy year-round tropical climates. The highland varieties will often experience colder conditions, sometimes with dramatic variations between day and night-time temperatures. Virtually all *Nepenthes* live in wet, high-humidity areas. Most inhabit rainforested localities, although a few species can be found in more open grassy regions. Most *Nepenthes* are terrestrial (rooted into the ground) and they will ascend, vine-like, up among the surrounding trees or shrubs. A few varieties are primarily epiphytic, dwelling among the branches of trees. Some species can reach 30 to 40 metres (98 to 131 ft) in length; others are much more compact.

As a new pitcher develops, it will swell – rather like a balloon – but its lid will remain sealed. Inside, fluid will begin to form. As with butterworts, some antibacterial qualities in the fluid of unopened *Nepenthes* pitchers have been identified, which probably help deter bacterial growth in the pitcher fluid.[30] When the pitcher approaches full size, the lid will slowly open. Contrary to popular belief, a pitcher lid will not close again, but will remain permanently open. The plant will continue to secrete fluid into the pitchers; they will also collect rainwater.

The pitchers attract their prey by a number of different means; these include bright coloration, nectar and, in some cases, scent. The traps are often described as comprising three different zones of insect capture: the attraction zone, the pitfall zone and the digestion zone.[31] The lid and the rim (peristome) of the pitcher opening,

Nepenthes maxima (upper pitcher), from around the Anggi Lakes,
West Papua, New Guinea.

areas covered in glands that exude a sweet sugary nectar, are referred
to as the attraction zone. Because of the slippery surfaces, insects
feeding on this nectar may quite often lose their footing and fall into
the traps. The pitfall zone is the upper part of the pitcher interior,
immediately below the rim. The walls of this part of the pitcher are

very waxy, making it nearly impossible for an insect to secure a foothold. The lower part of the pitcher is referred to as the digestive zone. In addition to digestive enzymes, the pitcher fluid also contains a type of wetting agent, which ensures that the insects will drown quickly. The digestive enzymes normally work rapidly and the soft

Nepenthes maxima (lower pitcher) from around the Anggi Lakes,
West Papua, New Guinea.

parts of most insects will become digested in just a couple days. Some pitchers contain a chemical known as chitinase which can digest even the most rigid exoskeleton of an insect. Many pitchers will also contain microorganisms and other small insects that will aid in the breakdown of larger food particles.[32]

Prior to the discovery of their carnivorous tendencies, many once thought that the primary purpose of a pitcher was to provide the plant with its own water reserves to utilize during times of drought. Of course, it is now known that their main function is for feeding and digestion; however, some pitcher plants that grow in Borneo do experience extremely wet seasons followed by very dry periods, and rainwater-filled pitchers will help to sustain the plants until the wet season returns.[33]

One of the most notorious pitcher plants, and perhaps the most menacing in appearance, is the fanged pitcher, or N. *bicalcarata* (derived from the Latin referring to 'two-spurs'). As the name suggests, this pitcher has sinister-looking fangs that project down from the base of its lid. The snake-like fangs secrete copious amounts of nectar, which accumulate at the tips. Insects, primarily ants, will feed off these, invariably losing their footing and falling into the fluid-filled pitchers. Ants are probably the most common food for most *Nepenthes*, although spiders and centipedes are frequently consumed and, occasionally, larger creatures such as frogs. It is not uncommon for birds to fish trapped insects out of the pitchers, or for rats and other mammals to look for a drink of water. And, on very rare occasions, these opportunists might end up drowning in the pitcher, becoming an extra-large meal for the plant.

Albany Pitcher Plant (*Cephalotus follicularis*)

The unique Australian pitcher plant *Cephalotus follicularis* is known as the West Australian pitcher or, more specifically, the Albany pitcher. Its natural habitat is highly localized, spanning just one short coastal stretch near the city of Albany in the state of Western Australia.

Albany pitcher plant (*Cephalotus follicularis*).

Although somewhat resembling *Nepenthes* in their pitcher forms, *Cephalotus follicularis* is classified quite separately and represents the sole species of the plant family Cephalotaceae.

The Albany pitcher plant produces two types of leaves: flat non-carnivorous ones (for photosynthesis) and pitcher-shaped leaves to eat insects. The pitchers are quite small, rarely exceeding 5 cm in length. Each has a lid and, although they do not snap shut, they will sometimes partly close (very slowly contract) in times of drought. The pitchers, which range from green to bright red in colour, are attached at the back to a stem and normally rest on the ground in order to attract crawling insects, primarily ants. The plants like full sun, and the more they are exposed, the redder their coloration will become.

Cephalotus follicularis in a coastal habitat in Western Australia.

Around the rim of the pitchers are curved, inward-pointing spines. Nectar is secreted in this area. An insect feeding on the nectar will usually slip and fall into the traps, each of which has three vertical ridges covered with hairs on the outside walls. These ridges guide the insects, usually crawling ants, up to the nectar glands on the pitchers' rim.

Like most carnivorous plants, their floral displays grow upon relatively tall stems, elevating them high above their carnivorous leaves. In the case of the Albany pitcher plant, their inflorescence can reach up to 60 cm (24 in.) in height.

North American Pitcher Plant (*Sarracenia*)

The genus *Sarracenia* comprises eight different species of pitcher plants, together with a number of subspecies and natural hybrids. Most of these feature tall, upright tubular pitchers with an umbrella-like lid. They are rather different in structure to *Nepenthes* in that, rather than being formed from the ballooning of a leaf tip, *Sarracenia* comprise a long hollowed-out leaf forming pitchers that can range from about 10 cm to 1.2 metres (4 to 47 in.) tall. Some varieties of pitchers can be bright and vividly coloured. Some people (and some rather unfortunate insects) have mistaken these unusual leaves for the plants' floral displays.

The plants are native to the eastern part of North America, although various species range from as far south as Florida, others as far north as northwestern Canada. The highest concentration of species, however, is in the southeastern United States. *Sarracenia* normally grow in constantly wet soils that are deficient in nutrients. They also thrive in direct sunlight; and those plants which grow in full sun will tend to have much more brightly coloured pitchers. In winter, when temperatures will often drop well below freezing, the plants die back, remaining dormant during this time.

Beginning in early spring, the plant will send up its tall flower stalks. Reaching its full height, the top part of the stalk will form a U-turn, causing the newly opened flower to dangle upside down, its petals hanging limply. Because of this unusual design, these pitcher plants are colloquially referred to as 'side-saddle flowers', reminiscent of the design of the traditional equestrian side-saddle, with the saddle flaps (and the rider's legs) tending to swing downwards.

The eight distinct species of *Sarracenia* comprise *S. alata* (pale pitcher plant), *S. flava* (yellow trumpet pitcher plant), *S. leucophylla* (white trumpet pitcher plant), *S. minor* (hooded pitcher plant), *S. oreophila* (green trumpet pitcher plant), *S. psittacina* (parrot pitcher plant),

Overleaf: Various *Sarracenia* cultivars.

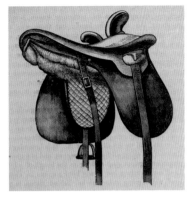

Side-saddle flower, a common name for *Sarracenia*, named as such because the flowers are thought to resemble the form of an equestrian side-saddle.

S. purpurea (purple pitcher plant) and *S. rubra* (sweet trumpet plant). Each of these eight recognized species also include a number of sub-species and further varieties. Thus a single species of pitcher plant might exhibit a very diverse range of colours, shapes and sizes. Particularly in the species *S. flava*, *S. rubra* and *S. leucophylla* a vast range of colours (white, greens, reds, purples, yellows) and patterns (stripes, flecks, splotches, filigrees) can be found within the pitchers. These variations, compounded with an almost countless range of hybrids, represent such stunning splendour and diversity that these pitcher leaves could arguably rival the flowers of tulips and roses.

Insects are attracted by the brightly coloured pitchers, many of which feature translucent areas that let through light in the manner of a stained-glass window. These 'illuminated windows' help to make the pitchers more attractive to approaching insects. The pitchers also exude a sweet nectar around the rim or lid of the pitcher. An insect that lands on this area and begins feeding will often lose its footing and fall in. In addition to having very slippery walls, the interiors of many pitchers are coated with downward pointing hairs, which together make it impossible for a fallen insect to climb out again. Some *Sarracencia*, such as the *Sarracenia flava*, lace their nectar with the drug coniine. This poisonous compound weakens and disorients the insect, often facilitating its fall into the pitcher.

The pitchers of *Sarracenia* are usually filled with a mixture of collected rainwater and digestive fluids that have been secreted by the plants into the pitcher wells. Insects will drown in this fluid, then be slowly digested. Some species, such as the purple pitcher plant (*S. purpurea*), feature rather short squat pitchers without lids. These are full of rainwater, will contain little or no digestive fluids,

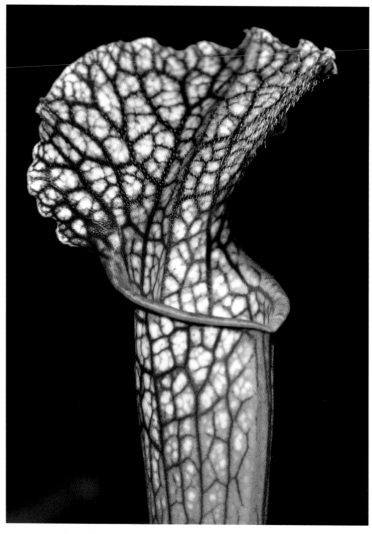

Sarracenia cultivar, known as fireworks, features an explosion of colour and pattern.

and will therefore rely primarily on other organisms to do the digestion for them.

Cobra Plant (*Darlingtonia californica*)

The genus *Darlingtonia*, commonly known as cobra plants, or the cobra lily (although they have nothing to do with lilies), contains just one species – *Darlingtonia californica*. As the name implies, it is native to California (specifically northern California and southern Oregon). It is closely related to the *Sarracenia* pitcher plants, belonging to the same family: Sarraceniaceae.

These plants have snake-shaped pitchers with odd-looking flaps reminiscent of the forked tongue of a cobra. This 'tongue' features nectar-producing glands that readily attract insects. An insect landing on the cobra's tongue will then follow the trail of sweet nectar up through the pitcher's narrow opening. Once inside the pitcher, it finds itself in a large, brightly lit domed space.

The roof and sides of the dome feature translucent 'windows' that allow dazzling coloured lights to shine in. This coloured light spectacle serves to disorient the captured insect so that, rather than trying to exit through the very narrow dark opening through which it came, it will attempt to fly upwards towards the lights. The walls of the cobra plant pitchers are very slippery and have downward pointing hairs that make it all but impossible for a captured insect to gain footing. Eventually, thoroughly exhausted and disorientated, the insect will tumble down into the pitcher's fluid.

Unlike most other North American pitcher plants, which have wide open pitchers, the watertight hood of the cobra plant does not allow in any rainwater. Instead, it produces and secretes all its own liquid into the pitchers, thereby being able to regulate carefully how much liquid each pitcher contains. This liquid does not contain any of its own digestive enzymes; the plant relies upon bacteria that have taken up residence within the water-filled pitchers to digest and break down its collection of the drowned insects.

Darlingtonia californica growing in its native habitat of northern California.

Similar to other pitcher plants in the Sarraceniaceae family, the flowers of *Darlingtonia* are distinctive in the manner by which they hang upside down in a side-saddle design. Cobra plants will also go into dormancy, dying back in winter. It propagates both by seed and, quite frequently, through underground runners which are sent out to form new clumps of genetically identical pitchers some distance away.

Sun Pitcher Plants (*Heliamphora*)

Sun pitchers, or *Heliamphora*, also belong to the family Sarraceniaceae. There are currently 24 species of sun pitchers in the genus, native exclusively to the high elevation, flat-topped mountains of South America known as tepuis. The plant produces pitchers that look somewhat like those of *Sarracenia* but which lack a developed lid. Most of them (like the cobra plant) do not produce their own digestive enzymes, instead employing microorganisms to do most of their digestive work.

The plants grow in environments that receive almost continuous rainfall. Thus, the pitchers are nearly always full of rainwater. However, to ensure that they don't overflow, and so as to not lose their valuable captured prey, most have small drainage holes near the top of the pitcher. These are covered by small 'filter' hairs that will let

Sun pitchers (*Heliamphora*), with characteristic nectar spoons visible on the rim of the pitchers.

the excess water out but keep the insects in.[34] In addition to heavy rainfall, most live in very sunny localities – thus the suitable common name of 'sun pitcher'.

The pitchers feature a small 'nectar spoon' in the place where other pitchers (such as *Sarracenia*) have lids. This nectar spoon is simply a small protrusion that secretes a sweet nectar. Insects are attracted to it and will land upon it to feed; but, because it is quite slippery, they will invariably fall into the water-filled pitchers. Upon drowning, they will be dissolved, with the help of resident bacteria and other micro-fauna. The freshly digested nutrients are then absorbed into the walls of the pitchers. Some species have pitchers that can reach to a height of 45 cm (18 in.). Most have varied colouring, including green, orange and red. Like other pitchers, they possess underground rhizomes that allow them to multiply freely.

Carnivorous Bromeliads

There are just a few species of bromeliads (within the larger family Bromeliaceae) that have been identified as being carnivorous. But although many bromeliads will collect rainwater in their funnel-shaped leaf bases, most are not carnivorous. Of the carnivorous species, two are from the genus *Brocchinia* (*B. hectioides* and *B. reducta*), native to the South American highlands of Guyana and Venezuela. They even look quite a lot like pitcher plants, with their form consisting mainly of upright leaf tubes.

Another bromeliad genus, *Catopsis*, includes just one carnivorous species: *C. berteroniana*. This species is primarily an epiphytic plant, growing almost exclusively among the branches of trees. The plants collect rainwater in their leaves, trapping insects and falling leaf matter. They also feature a covering of waxy white powder that helps to immobilize insects, facilitating their capture. They are tropical plants that have a wide distribution from South America up to the southern tip of Florida.

Corkscrew Plants (*Genlísea*)

Although not a pitcher plant per se, the remarkable *Genlísea* or cork-screw plant does feature numerous, very tiny, tube-like, subterranean leaves that are able to capture prey.[35] There are approximately thirty different species of corkscrew plant. Native to South America and Africa, like most carnivorous plants they can be found in wet, nutrient-poor soil. They are closely related to the bladderworts (*Utricularia*).

From above the ground, the corkscrew plant looks very much like any other small rosette-forming plant with green leaves. And at first glance, one might wonder how it got the common name of 'cork-screw'. But underground, instead of roots, the plant features thin,

Carnivorous *Catopsis*, which are primarily epiphytic plants of the Bromeliaceae family, on display at the annual Victorian Carnivorous Plant Society show, 2016.

Carnivorous
Brocchinia reducta of
the Bromeliaceae
(bromeliad) family.

tightly twisted, chlorophyll-free, white leaves. These leaves, which look a lot like roots, grow down into the wet soil. They are twisted (corkscrew-like) and form a sort of tunnel. Very small insect creatures are attracted to these and will crawl inside them. Once inside, the insects are compelled to keep climbing upwards because of the very narrow, continually twisting passageway and the numerous upward-pointing bristles. Eventually they will reach a larger cavern. It is here that they will be digested and absorbed by the plant.

two

More than Just a Meal

utualistic associations between plants and animals can be found frequently in the natural world. Such beneficial partnerships extend into the world of carnivorous plants, where researchers are constantly discovering more of these constructive arrangements.

The most obvious (non-predatory) association that carnivorous plants have with insects is in pollination – it would be very counterproductive, to say the least, to eat one's pollinators. To ensure that this does not happen, often a plant's flowers will be strategically situated high up and far away from its carnivorous traps. This will have the added benefit of attracting bees and other airborne pollinators. In the case of the American pitcher plant (*Sarracenia*), the blooms will normally occur well before most of the new season's pitchers have formed.

But many other relationships have developed which specifically help carnivorous plants to be better carnivores. Carnivory, after all, is not something that comes easily to a lot of plants and they will gladly accept any help they can get. These intricate relationships can involve a wide range of animals, from single-celled protozoa to mammals, such as bats and tree shrews. Additionally, some carnivorous plants are carnivorous for only part of the time – behaving much like other plants during other phases of their cycles. Some plants will only pretend to be carnivorous; instead of eating their prey, they will merely play with their food.

Pinguicula (butterwort) in bloom.

Carnivorous Plant and Animal Associations

Digestion is one of the areas in which carnivorous plants most often need assistance. The breaking down of complex organic materials can require very diverse processes for each unique compound. Even humans rely upon other organisms (in our case the billions of bacteria that reside within our intestines) to do a portion of our own digestive work.

Some carnivorous plants do not produce any of their own digestive enzymes and rely entirely upon other organisms (bacteria,

Pitchers of *Nepenthes*, or tropical pitcher plant.

protozoa and insects) organisms to do the digesting for them. In these cases, the plant ends up essentially absorbing the excrement of these 'helpers'. Animal dung contains all the nutrients that the plant needs and, because it is already digested, it is easy for the plant to absorb. In return, the plants will gladly provide the creatures a safe haven, food supplies or whatever else they might need. Even the more vegetarian *Nepenthes* pitcher plants, those that consume leaf-matter that has fallen into their pitchers, appear to get a lot of help from insects and other organisms in the digestion of their supper.

There is one variety of carnivorous plant, *Roridula*, that very theatrically relies upon other creatures to do its digestion. Native to South Africa, *Roridula* is a sticky-leafed plant which Darwin identified to be carnivorous in the 1870s. But later, when it was found not to produce its own digestive fluids, some began to regard it as being not quite carnivorous – considering it to be instead a 'semi-carnivorous' plant, or even a 'pre-carnivorous' plant.

The leaves of the plant are covered in sticky hairs (similar to the sundew), but instead of producing dewdrops of mucilage, it manufactures a sticky resinous substance. This resinous fluid, unlike mucilage, is unable to transfer digestive enzymes, so the plant is unable to break down the trapped prey on its own.[1] It therefore relies upon an insect known as an 'assassin bug' (*Pameridea*) to devour and digest the other insects that it traps. Incredibly, assassin bugs are able to walk freely all over the *Roridula*'s sticky leaves without getting trapped. When it encounters a trapped insect, it will stab its sharp mouth parts into the creature and suck out its innards. It will then digest this material and excrete drops of clear faeces onto the leaves of the plant. In this digested faecal form, the plant can easily absorb the nutrients. Rather than being described as a carnivorous plant, *Roridula* could be more accurately described as a coprophagous (dung-eating) plant.

The rainbow plant (*Byblis*) engages in a similar partnership, in this case using the capsid bug (*Setocoris bybliphilus*).[2] In addition, a number of species of Australian sundews (including *Drosera pallida* and *D. erythrorhiza*) rely upon insects, Cyrtopeltis bugs, to help with their digestion. These special bugs can also walk across the sticky leaves without fear of getting trapped.[3]

Some of the most intricate animal associations occur in conjunction with pitcher plants. Although most do produce at least a portion of their own digestive enzymes, some species, such as the cobra plant (*Darlingtonia*), rely entirely on third-party microorganisms

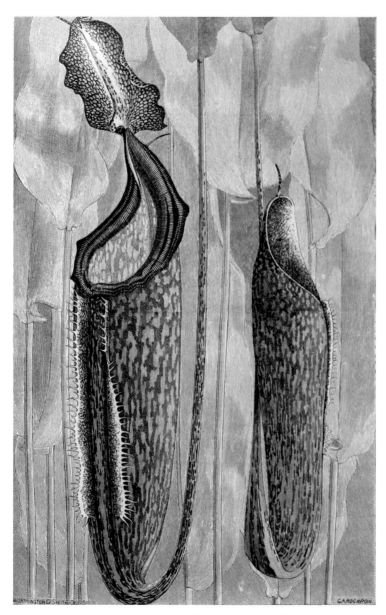

Nepenthes curtisii, from *L'Illustration Horticole* (1888).

Assorted *Sarracenia* pitcher plants.

which reside within their pitchers. But unlike most pitcher plants, *Darlingtonia* have watertight hoods over their pitchers, protecting them from rainfall. The plants therefore produce all their fluids themselves and, because of this, are able to regulate their own levels of pitcher fluid. This also gives them the unique ability to control how potent or diluted their digestive infauna might be.

Similarly, most species of sun pitchers (*Heliamphora*) rely almost entirely on resident micro-fauna. This seems particularly appropriate in these species because their pitchers do not have lids and, living in very wet climates, are continually overflowing with rainwater. It would

The long, green upper pitchers of *Nepenthes albomarginata*.

become very expensive for these plants to have to keep producing enzymes; it is much more efficient to contract out this work. The purple pitcher plant (*Sarracenia purpurea*), which also has very wide open pitchers that are constantly diluted with rainwater, is another species that relies upon microorganisms to do much of its digestion.

Fly and mosquito larvae are also commonly found within pitcher plants. The larvae of the pitcher plant mosquito, *Wyeomyia smithii*, are frequently found in *Sarracenia* and, remarkably, they seem to be completely resistant to this plant's highly acidic digestive fluids. Similarly, the larvae of *Sarcophaga* flies thrive within these pitcher fluids.[4] Many opportunistic spiders will reside just inside the pitcher plants of both *Sarracenia* and *Nepenthes*, taking advantage of the steady stream of prey. The pitchers still benefit as the spider will, after sucking out the insect's insides, invariably drop the remains into the pitchers below. Furthermore, the pitchers will also benefit from the nutrient-rich excrement of the spiders.[5] In Southeast Asia, over a hundred different species of animal life have been found living inside the pitchers of the various species of *Nepenthes*. Amazingly, many of these creatures live exclusively within the micro-environments of the carnivorous pitchers and cannot be found anywhere else.[6]

Ants make up the majority of insect prey for *Nepenthes*. However, one species of ant, the golden ant (*Camponotus schmitzi*), has formed an amazing and intricate association with the fanged pitcher, *Nepenthes bicalcarata*.[7] These ants live burrowed inside the stems of the pitchers (the tendril connecting the conventional leaf form to the pitcher) and seem to be able, effortlessly, to navigate the pitchers' very slippery walls and rim. These ants are amphibious, in that they are able to swim into the liquid of the pitchers and retrieve and consume insects that have drowned. The plant benefits because the ants keep in check the number of putrefying insects, which if too high might generate the risk of disrupting the plant's digestive system. As with other resident insects, the ants will also deposit their faeces into the pitchers, providing highly potent pre-digested nutrients.[8] Another very beneficial service that these ants provide is that of acting as sentinels, guarding against the most destructive pest that the *Nepenthes* must face – the plant-eating weevil *Alcidodes*. Whenever one of these is discovered on the plant, the ants will immediately attack it; but, surprisingly, they will completely ignore all other visiting insects.[9]

Another species of ant, a large black ant known as the drummer ant (*Polyrhachis pruinosa*), will frequently visit the pitchers and attempt to drink the nectar that forms on the sharp fangs that protrude down from the lids. Once they do this, they will invariably lose their footing on these slippery fangs and fall into the pitcher. This species of ant constitutes the main food supply for the pitchers (and, as a result, that of the resident golden ants as well).[10]

There are a number of species of frog that either live within or lay their eggs inside the *Nepenthes* pitchers. For example, a small frog known as the leaf-litter frog (*Philautus* sp.) will lay its eggs within the pitchers of several species, including: *N. bicalcarata* and *N. hirsuta*. But in order to protect its tadpoles from the acidic water of these pitchers, it lays just a few exceptionally large eggs. Unlike most frogs, the tadpole will not hatch until it has fully developed into a froglet, thereby avoiding having to swim in the acidic water. Once hatched, the fully formed frog will quickly hop out of the pitcher.[11] Another type of frog (*Microhyla* sp.) seems to have developed some resistance to the pitcher fluids. It lays its eggs into the pitchers of *N. ampullaria*, its young hatching and living out their term as free-swimming tadpoles within the pitchers. Once transformed into small frogs they will hop out, leaving the pitchers, only returning when they are ready to lay eggs of their own.[12]

Frogs are also often found within the purple pitcher (*Sarracenia purpurea*) and the sun pitcher (*Heliamphora*). Both of these have little or no digestive enzymes and are therefore primarily composed of rainwater – making a perfect waterhole in which a frog can hang out awaiting the arrival of insect prey. As in other instances, these plants probably don't mind if the frogs steal their food because, in doing so, they will accomplish the digestion for them, depositing their dung into the pitchers. The frog, of course, enjoys a comfortable and safe place to reside, as well as a steady supply of insect food.

Another remarkable association occurs between several species of *Nepenthes* (especially *N. lowii* and *N. rajah*) and mammals, particularly the tree shrew (*Tupaia montana*) and the summit rat (*Rattus baluensis*).

The pitchers of these plants provide the tree shrews and rats with nectar, and in return the animals use the pitchers as a toilet. *Nepenthes rajah*, in particular, has enormously large pitchers and the animal will stand on the rim of the pitcher and drink its nectar. In this position, the animal's bottom is placed directly over the mouth of the pitcher, and almost immediately upon commencing to drink the nectar the animal will begin to defecate. It has been observed that the tree shrews will visit during the day and the rats at night. In this way, the nectar

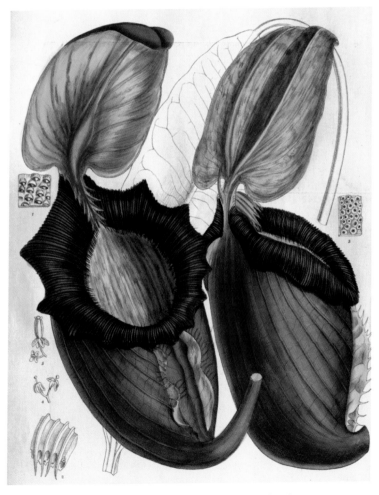

Nepenthes rajah, from *Curtis's Botanical Magazine* (1905).

Syd. Edwards del. Pub. by T.Curtis, St Geo: Crescent July 1.1805. F.Sansom sculp.

Purple pitcher plant (*Sarracenia purpurea*), from *Curtis's Botanical Magazine* (1849).

supplies are fresh for each species, and the pitcher can rely on a steady flow of manure both night and day.[13]

Birds are also attracted to the nectar of these pitchers and the mountain blackeye bird of Borneo has also been seen perched on the rim of the pitcher of *Nepenthes lowii* and similarly leaving its droppings into the pitcher fluid below.[14] Of course, not all interactions between these plants and animals are wholly mutualistic. In many instances,

A rat (*Rattus baluensis*) visiting a *Nepenthes rajah*.

Nepenthes curtisii from *Curtis's Botanical Magazine* (1890).

birds can be seen dipping into the pitchers and stealing insects from within. Yet, even in these cases, the birds will probably tend to remove the larger insects that the plant might have found difficult to digest.

In Borneo, there is a truly remarkable mutualistic relationship that has formed between a species of bat known as Hardwicke's woolly bat (*Kerivoula hardwickii*) and the tropical pitcher plant *Nepenthes hemsleyana*. The bat sleeps inside the pitchers of this plant, which seem almost specifically designed for these small flying creatures. The pitcher's interior includes a ledge-like formation that prevents the bat from sliding too far down into the pitcher. The pitchers also tend to contain much less fluid and are longer in length than many other *Nepenthes*, ensuring that the bats avoid getting wet. The bats also rarely have to contend with insects as these pitchers produce little or no nectar and are very ineffective at attracting them.[15] The pitchers are roomy enough for, in some cases, a parent bat to snuggle up with one of its pups. The pitchers offer the bat a safe place to sleep, which is generally free of parasites and gives protection from any extremes in the weather. In return for this, the bat defecates the digested remains of its night-time foraging directly into the pitchers, which the plant readily absorbs. It is estimated that nearly half of the plant's nutrients come from bat guano.

Even more remarkable is that these bats are able to locate, often in complete darkness, a vacant pitcher among the very dense foliage of the forest. To further complicate things, there are numerous similar-looking species of pitchers, which are by far more prevalent in the area. The bat, however, is able to locate their specific pitcher of choice, the *Nepenthes hemsleyana*, through echolocation. The bat will call out with a screeching sound and listen for the correct sound signature to return. Remarkably, the back wall of this particular pitcher plant 'features a prolonged concave structure, which distinctively reflects the bats' echolocation calls for a wide range of angles'.[16] Only this particular species of pitcher can send back the correct sonar wave-form signature, and do so with such a broad transmission that the bats are able to pick it up over a great distance. On rare occasions, probably when there is a scarcity of *Nepenthes hemsleyana*, the bats can be found sleeping inside other pitchers; but only in those in which the fluid has been drained out – namely those that have been damaged or are dying.[17]

Omnivorous Plants

Many carnivorous plants could be more accurately described as being omnivorous, as they will eat not only insects but a wide variety of animal dung and, in addition, they have frequently been found to eat

Pinguicula moranensis, native to Mexico, with a number of small insects
trapped on its sticky leaves.

vegetative material. It was Charles Darwin who first made note of the
fact that some carnivorous plants are also, at least partly, herbivorous
in their diet. He made special note of sundews: 'They likewise absorb
matter from pollen, and from fresh leaves . . . *Drosera* is properly an
insectivorous plant; but as pollen cannot fail to be often blown on to
the glands, as will occasionally the seeds and leaves of surrounding
plants, *Drosera* is, to a certain extent, a vegetable-feeder.'[18] He made
similar observations regarding the butterwort:

> We may therefore conclude that *Pinguicula vulgaris*, with its
> small roots, is not only supported to a large extent by the
> extraordinary number of insects which it habitually captures,
> but likewise draws some nourishment from the pollen, leaves,
> and seeds of other plants which often adhere to its leaves, it
> is therefore partly a vegetable as well as an animal feeder.[19]

On the same subject, Darwin's contemporary, the naturalist Andrew
Wilson, made note that this eating habit should be considered not
in terms of vegetarianism, but in terms of cannibalism: 'Lastly, it has
been noted that fragments of plant-tissue and pollen-grains are
also found on the leaves of the butterwort, and that, cannibal-like,

the *Pinguicula* may therefore devour parts of its neighbour-plants.'[20] Interestingly, Wilson seems to have been the first to have correctly used the term 'cannibal' when referring to carnivorous plants. In subsequent decades, as imaginative rumours of man-eating trees began to circulate, these fictional plants were also (yet inaccurately) referred

Pinguicula grandiflora.

W. Miller delt   G. C. sc.

William Miller, drawing of *Pinguicula grandiflora*, engraved by G. Cooke, from *The Botanical Cabinet Consisting of Coloured Delineations of Plants from all Countries* (1818).

Nepenthes ampullaria, native to Sumatra, a carnivorous plant
(but a vegetarian most of the time).

to as 'cannibals' – not because they ate other trees of the same species
(which would have been the proper use of the term), but because they
supposedly ate humans.

More recently it has been shown that several species of *Nepenthes*,
in particular the ground-dwelling *Nepenthes ampullaria*, have become
primarily herbivorous. This plant has very wide open pitchers, does
not appear to produce nectar, and will obtain most of its nitrogen
needs from leaf-litter that has fallen into its pitchers from the forest
canopy.[21] In contrast, the Venus flytrap is perhaps one of the most
carnivorous of the carnivores. As one expert has pointed out, because
of the need for mechanical stimulation – that is, the food needs to
be alive and moving in order for the plant to begin to secrete digestive
enzymes and begin feeding – it would be fairly unlikely for Venus
flytraps to gain much nutritional benefit from vegetable matter.[22]

But regardless of what a carnivorous plant eats, it is how they go
about doing so that clearly sets these plants apart from most others.

It certainly appears, at least from our human perspective, that they possess a clear capability (perhaps even an intention) to lure, capture, digest and consume their food through their specialized leaves, be it insects, small animals, pollen, leaf-matter or animal dung.

Part-time Carnivores

Some carnivorous plants, however, are carnivorous for only part of their lives. Being carnivorous takes a lot of effort, and a plant needs to get a big enough reward to make this expenditure worthwhile. Many carnivorous plants will go dormant in winter, not only because their metabolism becomes less efficient in cold weather, but from the point of view of insect availability. Quite simply, there are more insects around in warmer weather. Many species of butterwort, particularly those native to Mexico such as *Pinguicula debbertiana*, *P. macrophylla* and *P. gypsicola*, will swap their normal carnivorous leaves for much thicker succulent leaves in wintertime. Such a strategy makes sense, because these species of butterwort grow in very dry habitats and these succulent carnivores can often be found growing alongside cacti and other desert plants.[23]

There are also a couple of carnivorous plants that are carnivorous only during particular stages of their lives. One of these is the Australian sundew, *Drosera caduca*. In the first stages of its life, it grows sticky, insect-trapping leaves. But as it gets older it stops producing carnivorous leaves and transforms into a 'normal' plant in its old age.[24]

Even more remarkable in this regard is the *Triphyophyllum peltatum*. This amazing plant, which is native to the West African rainforests, goes through three distinct stages of life, only one of which is carnivorous. As a young plant, it grows medium-sized leaves on a woody stem up to a metre tall. After a few years, it will enter its second phase of life – its carnivorous stage. At this point its leaves become covered in glandular hairs like a sundew or a dewy pine and it becomes fully carnivorous, capable of digesting it prey through its leaves. In its final stage of life, it loses these carnivorous leaves and then, in a remarkable

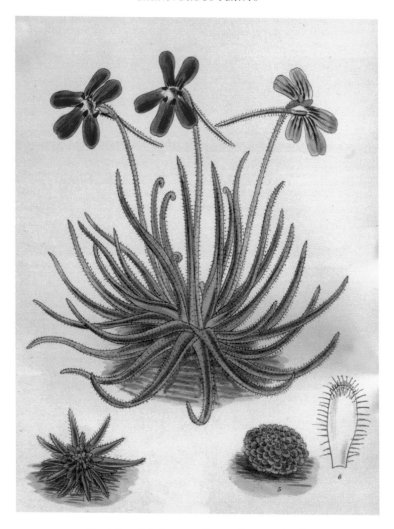

Pinguicula gypsicola, from *Curtis's Botanical Magazine* (1915).

expenditure of saved-up energy, grows into a vine-like climbing plant, rising to over 45 metres (148 ft) and into the forest canopy. Once it reaches the tops of the trees, it produces its flowers and then its seeds are dispersed by the wind to distant parts of the forest.[25]

Playing with their Food

There are a number of plants that appear to be only partly carnivorous. Such plants will often trap and kill insects but not actually eat them. Some have classified these as being semi-carnivorous or pseudo-carnivorous, often with the suggestion that they are on an evolutionary trajectory towards full carnivory. However, some species have been identified as acting primarily in a defensive mode; that is, they will actively deter and even kill insects and other creatures in order to keep them away from their flowers or to prevent them from eating their leaves. Before their carnivorous nature was proven, some believed that the Venus flytrap caught insects primarily as a defensive measure – a sort of 'get them before they get us' approach. Sundews and pitcher plants were also suspected of engaging in such defensive behaviour; some even suggested that they could be playing a role in Nature's grand design in order to keep the world's insect population in check. More recently, some writers have, with a bit of tongue-in-cheek anthropomorphism, described these plants as

Sarracenia psittacina, commonly referred to as the parrot pitcher plant.

being 'murderous' because they will kill not just for sustenance, but for perhaps less ethical reasons.[26]

There are several non-carnivorous plants that will often trap and kill insects within deep pools of water. Many bromeliads, for example, will collect puddles of rainwater in the leaves at their base. Countless insects, both living and dead, can be found within these pools. Yet so far, only three species of bromeliad have been identified as being truly carnivorous. Some agree, however, that even the non-carnivorous bromeliads probably do benefit, albeit in minute ways, from the decaying animals that might have drowned in their pools. It is possible that this slight benefit may provide just enough of an advantage to, over thousands of years, nudge a plant towards becoming fully carnivorous.[27]

Many plants (both carnivorous and non-carnivorous) will trap insects upon their glandular sticky hairs, but one variety that warrants special note is the Australian trigger plant (*Stylidium*). What makes this genus of plants so intriguing is how the plants will engage with insects in two very different ways. First, the plant will catch copious numbers of small insects in the minute sticky hairs that cover its leaves and stems. Although this certainly resembles the actions of a carnivorous plant, their carnivory has yet to be proven. Second, and more remarkable, is the manner in which they deal with their pollinator insects, through their extraordinary capacity for fast movement. Rather than using their speed in a lethal manner to trap and kill insects, as does the Venus flytrap, or in a defensive manner, as does the 'sensitive plant' *Mimosa pudica*, these plants use it to facilitate cross-fertilization with their pollinating insects. When an insect lands on its flower, a trigger arm will (within a few milliseconds) snap forward and violently smash a mound of pollen onto the back of the insect, completely covering it. The insect will then unwittingly carry this to the next flower.

Because of the sticky hairs which cover the trigger plant's stems, some of which have shown evidence of producing enzymes, these plants

Nepenthes bicalcarata with its fangs visible on the underneath side of its lid.

have long been suspected of being carnivorous. Another factor that has encouraged carnivorous suspicions of these plants is that they almost always grow in nutrient-poor soil, alongside *Drosera* and other carnivores. Of course, there is no such thing as being 'carnivorous by association', but these details have provided a compelling motivation for further study.[28]

three
A Remarkable Discovery
Cᶜᴰᴺᵃ

The discovery of carnivorous plants was one that seemed to up-end the world view of a great many people, but it did not happen overnight. Although most of these plants were well-known long before their carnivorous nature was defined, it took quite some time for their carnivority to be generally accepted.

Many species were regarded as being quite extraordinary before they were known to be 'meat-eaters'. Some, for example, were thought to possess strong medicinal powers and in some cases were shrouded in mysticism. Butterworts (*Pinguicula*) had been used extensively in northern Europe for their antibacterial properties. The plant's sticky mucilage would be rubbed onto the wounds of sheep and cattle to fight against infection. *Nepenthes* pitcher plants were used in traditional medicine in Southeast Asia, with the fluid of unopened pitchers used to treat eye and skin irritations.[1] Some drank the fluid to relieve coughs and breathing problems.[2] Others believed that the pitchers held miraculous properties. The Iban people of Borneo would regularly mix the burnt ashes of pitcher plants into the food they fed to their domesticated dogs and pigs as they believed it would protect these animals against snake and scorpion bites.[3]

When thirsty European explorers first encountered pitcher plants they were deemed to be nature's way of providing much needed drinking water. In 1686 John Ray wrote in his *Historia Plantarum* of *Nepenthes*:

An upper pitcher of a *Nepenthes*, characteristically facing away from its tendril, on display in Singapore.

These reservoirs of liquid, until fully mature, are closed by an elegant lid, which when pressed by a finger splits open to supply an abundance of clear, soft, pleasant, limpid, cool and refreshing water, so that from six to eight of these reservoirs contain enough water to quench the thirst of a single man very delightfully.[4]

The sundew was also considered to be an extraordinary plant, people marvelling at how its 'dew' seemed never to evaporate. In Henry Lyte's *A New Herball* (1578) he describes the sundew:

This herb is of a very strange nature and marvelous: for although that the sun does shine hot, and a long time thereon, yet you shall find it always moist and bedewed, and the hairs thereof always full of little drops of water: and the hotter the sun shineth upon this herb, so much the moistier it is and the more bedewed.[5]

And on the Isle of Man, the sundew was sometimes used as a love charm. The sticky leaves would be 'surreptitiously slipped into the clothing of the person who was to be attracted'.[6] Presumably, the person's heart would then be captured by the sticky leaves – and this love would subsequently never dry up.

Of course, the fact that pitcher plants tended to be full of putrefying insects, or that sundews would happen to have scores of small insects stuck to their glands, did not go completely unnoticed. However, these observations were largely dismissed as either unfortunate coincidences or as defensive measures. Some even speculated that it was part of nature's ingenious way of keeping the insect population in check.[7]

However, when people began to suspect that these plants might be carnivorous, it motivated some to question the overly rigid hierarchy that scientists and theologians had traditionally advocated. It was a hierarchy that stipulated that plants were indisputably lower in importance than animals. Nehemiah Grew was one of several renegade naturalists who, early on, led the way in suggesting that plant life was virtually the same as animal life. In his introduction to *The Anatomy of Plants* (1682) he wrote:

There are those things within a Plant, little less admirable, than within an Animal. That a plant, as well as an animal, is composed of several organical parts; some whereof may be called its bowels. That every plant hath bowels of diverse kinds, containing diverse kinds of liquors. That even a plant lives partly upon air; for the reception whereof, it hath those

Overleaf: Detail of a sundew, showing glistening dewdrops of sticky mucilage.

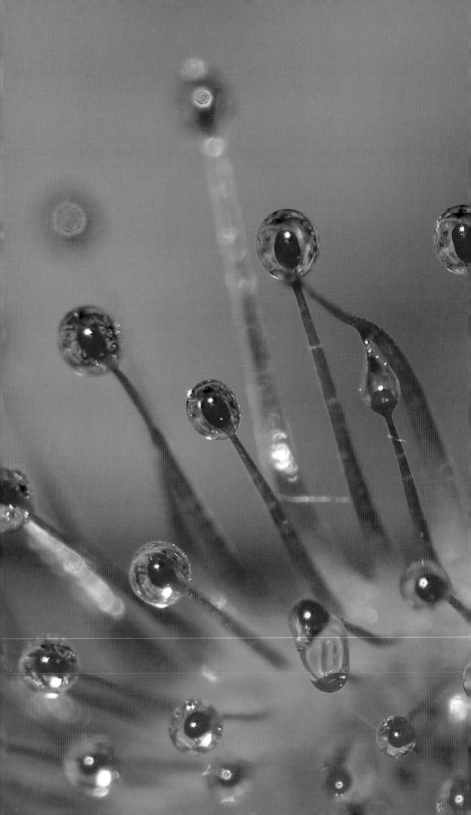

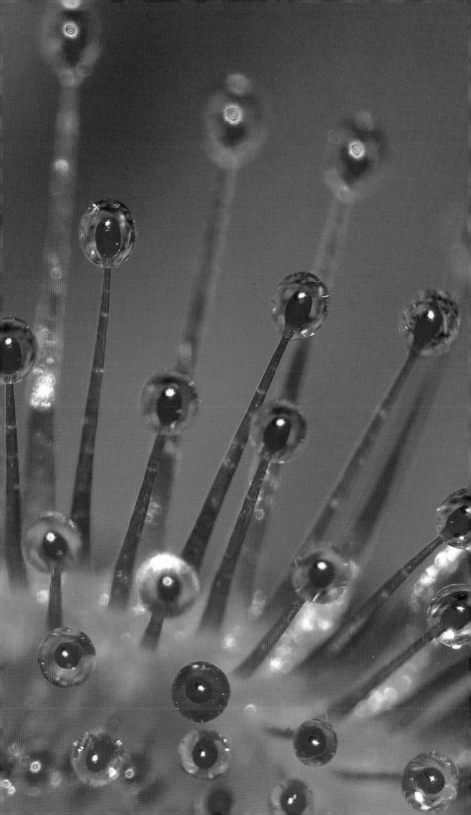

NEPENTHES VILLOSA *H.f*

Published by Smith, Elder &C°. 65 Cornhill London.

Botanical print of *Nepenthes villosa* from Spencer St. John,
Life in the Forests of the Far East (1863).

parts which are answerable to lungs. So that a plant is, as it were, an animal in quires; as an animal is a plant, or rather several plants bound up into one volume.[8]

To say that an animal was composed of 'several plants bound up into one volume' was a strong proclamation that, on a fundamental level, plants and animals were identical. It was only a matter of compounded complexity that differentiated the two. John Ellis, a naturalist who specialized in zoophytes (creatures, such as sea sponges, that were considered to have both animal and plant-like characteristics), leapt at the chance to study the Venus flytrap. He was the first to publish a detailed description of the plant, and the first to publically suggest that they consumed their insect prey. Charles Darwin also seemed to delight in exploring the boundaries of convention. His wife, Emma Darwin, in a letter dated 1860 to her friend Mary Lyell, gently teased her husband's efforts, writing: 'at present he is treating Drosera [sundews] just like a living creature, and I suppose he hopes to end in proving it to be an animal.'[9] A contemporary of Darwin's, T. H. Huxley, noted that there was an expanding 'border territory between the animal and vegetable kingdoms'. He therefore decided that 'to be on the safe side, I shall call such things as the sundew, Venus's flytrap, and so forth, vegetable animals.'[10]

Discovering the Venus Flytrap
(*Dionaea muscipula*)

The Venus flytrap (or Venus's flytrap as it was initially named) is native to a very localized area of North Carolina. It was first discovered in 1759 by Arthur Dobbs, who was Governor of the then British colony. He happened to stumble across the plant growing on his property and, witnessing its remarkable movement, hurriedly sent correspondence to his horticulturalist friend Peter Collinson in England. He wrote: 'I have taken a little plantation at the sound on the sea coast. We have a kind of a Catch Fly Sensitive which closes upon anything that touches it. It grows in the Latitude 34 but not in 35. I will try to save the seed here.'[11] Dobbs wrote to Collinson again, some months later, with a bit more detail:

> But the great wonder of the vegetable kingdom is a very curious unknown species of sensitives; it is a dwarf plant; the leaves are like a narrow segment of a sphere, consisting of two parts, like the cap of a spring purse, the concave part outwards, each of which falls back with indented edges (like an iron spring fox trap); upon any thing touching the leaves or falling between them, they instantly close like a spring trap, and confine any insect or any thing that falls between them; it bears a white flower: to this surprising plant I have given the name of Fly Trap Sensitive.[12]

Eventually, the plant was named *Dionaea muscipula*. The genera name of *Dionaea* derives from the ancient Greek goddess Dione, who was the mother of the goddess of love, Aphrodite. But, somewhat confusingly, the species name *muscipula* is translated as 'mousetrap', not the expected 'flytrap'. Its most regularly used common name, the Venus flytrap, refers to the Roman goddess. As Richard Mabey has noted of its early discovery, 'the name's curious, almost oxymoronic linking of the Goddess of Love with a spiked gin trap tickled public imagination even further, and letters, pamphlets and engravings flew about botanical circles'.[13]

In some of these correspondences that 'flew about botanical circles', the plant, with its showy red flytraps, was referred to by the obscure, but nonetheless bawdy, term 'tipitiwitchet' – said to be a reference to female genitalia.[14] The American botanist John Bartram, who also corresponded with Collinson, wrote to him on 29 August 1762: 'my little tipitiwitchet sensitive stimulates laughter in all ye beholders; there was lately a French gentleman from Montreal which was so agitated that he could hardly stand & said it was enough to make one burst with laughing'. In a reproachful reply, Collinson,

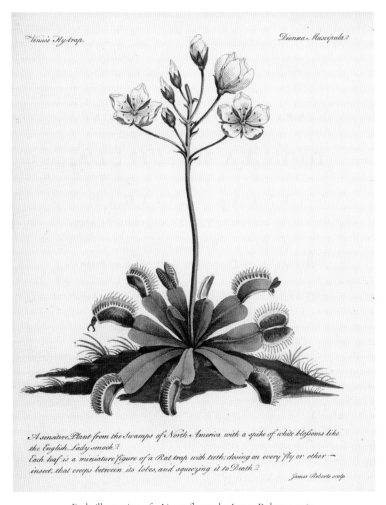

Early illustration of a Venus flytrap by James Roberts, 1769.

who had not yet seen the plant personally, wrote: 'Whilst the French Man was ready to burst with laughing I am ready to burst with desire for root, seed, or specimen of the Wagish Tipitwitched Sensitive.'[15]

One prominent person who was able to acquire several living specimens of the plant was the highly regarded English naturalist John Ellis. The plant's fast movement and unusual trapping mechanisms immediately fascinated him. Significantly, Ellis had previously published extensively on zoophytes and was, therefore, very interested in creatures that appeared to traverse the boundaries between the plant and the animal kingdoms. Based on his initial observations, he became the first to speculate that this strange plant might not be merely catching insects, but might also be eating them. In 1768, in his first published article on this plant, he articulated its scientific name as *Dionaea muscipula* and noted that the name of 'Venus's Flytrap . . . may be admitted to be the most eligible trivial name'. He also spelled out many of its remarkable characteristics and mentions the 'surprising faculty of its entrapping little animals, such as ear-wigs, spiders, and flies, where they are either squeezed to death, or remain imprisoned till they die'.[16]

As with Ellis, there were others who seemed keen to find common ground between the plant and animal kingdoms. After reading his article, William Logan Jr, a student from the University of Edinburgh, wrote to Ellis in 1769:

> Excuse me if in the vanity & fullness of my heart I once more enter on my own opinion viz. that there is a chain by which

Venus flytrap cultivars, displaying numerous hair-trigger snap-traps.

plants and animals are connected and that there is an
Amphibious State neither entire Plant nor Animal. We have
heard of Mermen & Mermaids. We have seen Sea Dogs &
Sea Lyons. We have the Bat and numerous Instances in other
parts of the Creation where the animal belongs to two classes.
Why not then in Plants? Let us reflect on the Dionaea,
Sensitive Plant & c. and enquire into the final end of their
Specific Differences and Phenomena![17]

Although Ellis probably regarded this query as rather naive with its
fanciful references to 'mermen and mermaids', he did nonetheless
continue to study the plant with the deep belief that, like an animal,
it could gain nutritional benefit from its bevy of captured insect prey.

In his 1770 publication, Ellis began with the use of some rather
appropriate culinary-themed language, 'I know that every discovery
in nature is a treat to you; but in this you will have a feast.' He then
continues:

> You have seen the Mimosa, or Sensitive Plants, close their
> leaves, and bend their joints, upon the leaf touch: and this
> has astonished you; but no end or design of nature has yet
> appeared to you from these surprising motions: they soon
> recover themselves again, and their leaves are expanded as
> before. But the plant, of which I now enclose you an exact
> figure, with a specimen of its leaves and blossoms, shows
> that nature may have some view towards its nourishment, in
> forming the upper joint of its leaf like a machine to catch
> food: upon the middle of this lies the bait for the unhappy
> insect that becomes its prey.[18]

Ellis corresponded with Carolus Linnaeus directly about his strong
suspicions that the Venus flytrap was carnivorous. Linnaeus, a staunch
traditionalist, quickly refuted this suggestion, claiming that such a
characteristic would be 'against the order of nature as willed by God'.[19]

Over the next hundred years or so there continued to be conflicting points of view as to whether or not these plants could be carnivorous. Erasmus Darwin (grandfather of Charles Darwin) suggested that these plants most probably killed insects as a defensive measure. He noted, for example, that the sticky glands of the sundew, 'prevents small insects from infesting the leaves, as the ear-wax in animals seems to be in part designed to prevent fleas and other insects from getting into their ears'.[20]

While the debate continued in scientific circles, the Venus flytrap with its highly visible insect-trapping activities and its thematic link to the 'goddess of love' had become an enduring subject of popular consciousness. One religious booklet from 1830, published by the American Tract Society, equated the workings of the Venus flytrap to broader themes of morality. The publication described how a flytrap will lure and tempt insects to its leaves but once 'an insect settles upon it, the [trap] closes together, and it cannot get away unless it is opened by some one'. This is comparable, the author explains, to 'sinful pleasures, which appear very tempting, and young persons think they may just try them a little; but when they have once done so, they are so entangled that they cannot escape unless freed by divine grace.'[21]

Discovering Pitcher Plants

From the other side of the world, another strange grouping of plants, tropical pitcher plants, had recently become known to Europeans. Although these plants had been well known for thousands of years by the many indigenous groups of Southeast Asia, they were first documented in publication in 1658. The then French governor of Madagascar, Étienne de Flacourt, described the pitcher plants in his text, *Histoire de la Grande Isle de Madagascar*, which also included what is probably the first published illustration of the plant. He referred to the species as *Anramatico* (now known specifically as *Nepenthes madagascariensis*):

It is a plant growing about 3 feet high which carries at the end of its leaves, which are 7 inches long, a hollow flower of fruit resembling a small vase, with its own lid, a wonderful sight. There are red ones and yellow ones, the yellow being the biggest. The inhabitants of this country are reluctant to pick the flowers, saying that if somebody does pick them in passing, it will not fail to rain that day. As to that, I and all the other Frenchmen did pick them, but it did not rain. After rain these flowers are full of water, each one containing a good half-glass.[22]

A few years later, another species was discovered in Sri Lanka and was identified as *bandura*, which was based on the local name. This plant is now known as *Nepenthes distillatoria*.[23] By 1737 Carolus Linnaeus, in his seminal publication *Species Plantarum*, formally adopted the name *Nepenthes*, thereby cementing its modern nomenclature.

The word 'Nepenthes' is of Greek origin and refers to an intoxicating substance that removes all grief and sorrow in those who drink it. Nepenthes is mentioned in Homer's *Odyssey*: 'She [Helen] threw a drug [nepenthes] into the wine, from which they drank that which frees men from grief, and from anger, and causes an oblivion of all ills.' In referencing this passage, Linnaeus noted, 'if this is not Helen's Nepenthes, it certainly will be for all botanists. What botanist would not be filled with admiration if, after a long journey, he should find this wonderful plant. In his astonishment past ills would be forgotten when beholding this admirable work of the Creator.'[24]

It is believed that Joseph Banks was the first to bring *Nepenthes* to England, introducing a specimen of *N. distillatoria* to the Royal Botanic Gardens at Kew in 1789.[25] During the following decades many other species of *Nepenthes* pitcher plants were discovered and slowly made their way into European collections.

Significantly, in the 1830s, Dr Nathaniel Bagshaw Ward invented the Wardian case. This was essentially a miniature, portable greenhouse — a case made of glass that allowed for the long-distance

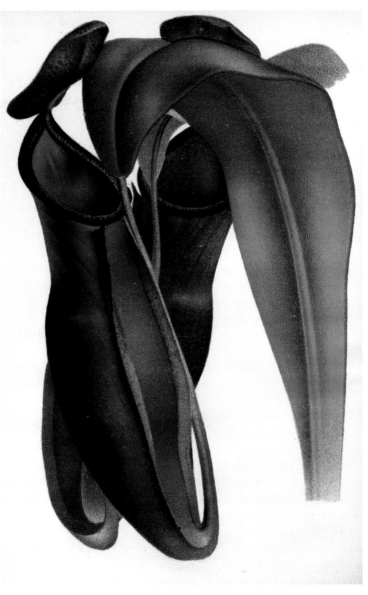

Nepenthes mastersiana, shown in an illustration from Charles Antoine Lemaire, *L'Illustration Horticole* (1886).

The lower pitcher of hybrid *Nepenthes justinae* × *peltata*.

transportation of fragile plant species. For the first time living *Nepenthes* (and many other exotic plants) could be shipped across the globe and would stand a very good chance of survival during the long sea voyage. By the 1860s pitcher plants were becoming a frequent addition to the collections of the wealthier plant enthusiasts throughout England, Europe and North America.

PLATE 1

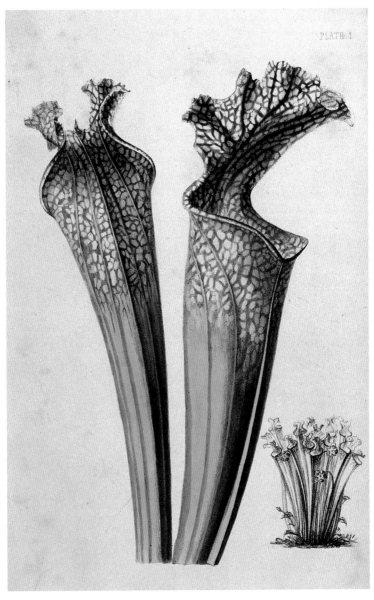

Illustration of *Sarracenia drummondii*, from John Lindley
and Joseph Paxton's *Paxton's Flower Garden* (1850).

Discovering Carnivority

Before it was known that pitcher plants (*Nepenthes* and *Sarracenia*) were carnivorous, there were various theories put forward to explain the function of their water-filled pitchers. Some suggested that these were simply for storing water for when the plants might have to endure dry conditions. Others speculated, upon seeing numerous live insects inside, that the pitchers might be designed as places of refuge for insects so that their animal predators could not eat them. It was further suggested that these were simply nature's way of providing thirsty birds and other animals with a drink of water.[26]

However, in J. E. Smith's *Introduction to Physiological and Systematic Botany* (1809) the author suggests that their pitchers are likely to be utilized for different purposes:

> But the consideration of some other species renders this hypothesis very doubtful. *Sarracenia flava* and *S. adunca* are so constructed that rain is nearly excluded from the hollow of their leaves, and yet that part contains water which seems to be secreted by the base of each leaf. What then is the purpose of this unusual contrivance? . . . The *S. purpurea* is usually observed to be stored with putrefying insects, whose scent is perceptible as we pass the plant in a garden; for the margin of its leaves is beset with inverted hairs, which, like the wires of a mouse-trap, render it very difficult for any unfortunate fly that has fallen into the watery tube, to crawl out again. Probably the air evolved by these dead flies may be beneficial to vegetation, and, as far as the plant is concerned, its curious construction may be designed to entrap them, while the water is provided to tempt as well as to retain them.[27]

Although he stops short of a full claim of carnivorousness, the author was one of the first to describe these pitchers to be a kind of trap. Rather than claiming that the pitchers digest and absorb nutrients from the insects, Smith makes a more moderate suggestion that perhaps the odours and gasses of the decaying insects might provide a more idyllic atmosphere for the plants to flourish in.

Darwin's Proof

Charles Darwin certainly wasn't the only one to study the carnivorous character of these plants, but he was the most famous. His book *Insectivorous Plants* (1875) was very widely read and made a huge impact both in the scientific worlds and among the general public. In the introduction, Darwin admits that he began his study with very little knowledge of the subject: 'During the summer of 1860, I was surprised by finding how large a number of insects were caught by the leaves of the common sun-dew (*Drosera rotundifolia*) on a heath in Sussex. I had heard that insects were thus caught, but knew nothing further on the subject.'[28]

But after ten years of dedicated research and many remarkable discoveries, he very quickly became the world's foremost expert on carnivorous plants. At one point, when still worried about how his research might be received, he wrote to his friend Charles Lyell proclaiming:

> At this moment, I care more about Drosera than the origin of all the species in the world. But I will not publish on Drosera till next year, for I am frightened and astounded at my results . . . Is it not curious that a plant should be far more sensitive to a touch than any nerve in the human body? Yet I am perfectly sure that this is true.[29]

Though much of *Insectivorous Plants* is devoted to his study of sundews (*Drosera*), Darwin did manage to study and report on most of the plants that were suspected of being carnivorous. The book also devotes an entire chapter each to the Venus flytrap, the waterwheel plant and butterworts, three chapters to bladderworts and

Illustration by Fritz Bergen, depicting *Pinguicula,*
Drosera and *Utricularia* in habitat, 1906.

A cut-away view of a *Nepenthes* pitcher, revealing an abundance of captured insects.

the corkscrew plant, and a final chapter to the sticky hair carnivores: the dewy pine and the rainbow plant.

In the book, Darwin also endeavoured to explain how such a diverse array of species could have evolved separately (and sometimes at different ends of the earth) towards a parallel form of carnivority, speculating:

> Any ordinary plant having viscid glands, which occasionally caught insects, might thus be converted under favourable circumstances into a species capable of true digestion. It

ceases, therefore, to be any great mystery how several genera of plants, in no way closely related together, have independently acquired this same power.[30]

Across the world at this time, other researchers were also experimenting with and theorizing about carnivorous plants. John Burdon Sanderson did methodical experimentation on the presence of electrical currents in plants, with particular emphasis on how electrical pulses are used to trigger the closure of the leaves of the Venus flytrap. He first presented his findings in 1874 and expanded upon his research in his book *Excitability of Plants* (1882). Mary Treat, a well-respected naturalist, devoted many years to the study of carnivorous plants, in particular *Sarracenia* and *Utricularia*. She detailed many of her observations in her best-selling book *Home Studies in Nature* (1885). Also researching carnivorous plants was the father and son team of William and Joseph Hooker – who together briefly occupied the joint role of director of the Royal Botanic Gardens, Kew. Joseph Hooker went on to publish 'The Carnivorous Habits of Plants' (1874), which focused primarily on the pitcher plants: *Nepenthes*, *Sarracenia* and *Darlingtonia*.[31]

After the publication of Charles Darwin's *Insectivorous Plants*, there emerged a general consensus that a class of plants that were carnivorous indeed existed. And as further publications were released, both by Darwin and by others, there were very few people left who would doubt they were real.

Discovering Vegetable Wickedness

When the true carnivorous nature of these plants became common knowledge, the public's attitude towards them seemed to shift dramatically. Almost overnight, these plants went from being mere curiosities to being regarded by many as horrible monsters.

In 1855 (prior to Darwin's publication), in her book *The Flowering Plants of Great Britain*, Anne Pratt speculated as to why so many insects would become trapped in the leaves of the sundew. In her musing, she acknowledges that there had been some speculation as to whether they might gain nutritional benefit from the insects. However, she regards this merely as an intriguing idea that has 'sometimes excited a smile', suggesting instead that the

> best reason to be assigned for [the sundew's] fatal power over the insect race would be, that it is in accordance with that law reigning throughout Nature, by which one kind of created things preys upon another, thus keeping the number of all within due limits, and preventing any serious departure to be made from that variety which gives to the earth one of its greatest charms.[32]

Therefore, from this conciliatory perspective, sundews were engaged in nothing sinister, but instead in the rather noble task of helping to maintain the balance of Nature.

However, soon after Darwin's publication a very different tone emerged, as is evident in J. G. Hunt's essay from 1882. In describing the sundew, he declares:

> I have seen the delicate and painted fly disjointed and crushed, its splendid and wonderful eyes torn asunder by the hungry plants; the little moths, themselves but velvet atoms rejoicing in the warm sunlight, but now slowly melting away, limb by limb, in these horrible vegetable stomachs; their life and marvellous form and startling beauty all wrecked to feed a sundew in one of these Droserian graveyards . . . Surely here, if anywhere, is *vegetable wickedness*.[33]

In his popular natural history and gardening book *Alpines and Bog-plants*, author Reginald Farrer takes a similar view. His book describes, in vivid detail, various species of plants and how to grow them. But when he comes to the sundew he dramatically breaks from his conventional writing style, and declares:

Evil little things they are, with their carnivorous habit. One
wonders what crime the past lives of Drosera can have held,
that now the race should be compelled to [endure] so ominous and unpleasant a [fate] of murder and fraud. When
will the Sundews be free of the burden, through some self-
sacrificing individual plant who shall starve to death rather
than take life, and so redeem his race into the happier paths
of peace and virtue?[34]

Then, as if remembering that he is supposed to be writing a book
about gardening, Farrer continues: 'Not to pursue such high inquiries
beyond what is fitting, I will merely add that the Sundews are not hard
to establish in wet moss.' But then, unable to resist adding one final
ridicule – this time directed towards the sundew's inflorescence – he
notes, 'their flower-spikes always promise much more than they
perform, only one or two blooms opening at a time . . . and never
producing any fine unanimous effect of blossom.'[35]

In a similar vein, *Colin Clout's Calendar: The Record of a Summer* (1883)
details how the author has frequently observed trapped insects on
the sundew's sticky leaves:

For my own part, I cannot watch the poor creature struggling
to free his legs and wings from this horrible, impassive, blood-
sucking plant without at once assisting him out of his trouble;
for my instincts will not allow me to appraise the 'divine
dexterity' of nature in causing destruction so highly as some
of our idealistic humanitarians have done; it is impossible
not to feel a little thrill of horror at this battle between the
sentient and the insentient, where the insentient always wins
– this combination of seeming cunning and apparent hunger
for blood on the part of a rooted, inanimate plant against a
breathing, flying, conscious insect.[36]

Overleaf: The forked sundew (*Drosera binata*).

Here Clout foreshadows what, in the coming decades, will become a recurring theme in the 'killer plant horror genre'. In these tales, the plants, almost zombie-like and without consciousness, will invariably 'hunger for blood'. However, rather hypocritically, the author continues, 'but with a little bit of raw beef one can see the whole process just as well, and far less cruelly' – implying, of course, that

Illustration showing pitcher, leaves and inflorescence of *Nepenthes*, published in *Curtis's Botanical Magazine* (1847).

the cow, in contrast to the 'conscious insect', would quite happily be consumed by a bloodthirsty carnivorous plant.

Perhaps most remarkably, in the revised 1884 edition of his book, the Italian criminologist Cesare Lombroso suggests that the human propensity to commit evil derives, almost entirely, from plants. To support his argument, Lombroso uses the existence of carnivorous plants as evidence: 'They [carnivorous plants] establish that premeditation, ambush, killing for greed, and, to a certain extent, decision-making (a refusal to kill insects that are too small) are derived completely from histology or the microstructure of organic tissue – and not from an alleged will.'[37] Clearly Lombroso was so astounded by Darwin's highly publicized depiction of carnivorous plants that he found it entirely feasible to advocate the shifting of blame from human will to some unknown potency found deep within the botanical make-up of carnivorous plants. (There is, however, no record from this period of anyone successfully avoiding criminal charges by diverting the blame to their Venus flytrap.)

Reporting on Carnivorous Plants

By the early 1900s, newspapers began to run regular stories about these unusual plants. Such articles seemed to delight in using very colourful and decidedly zoomorphic language in their descriptions, and quite often they would slip in some very inaccurate information. One remarkable story published in 1912 detailed a claim by Professor William Wallace Campbell (of Lick Observatory, California) that Mars was inhabited by intelligent plants. In order to justify his claim, he referred to the manner in which carnivorous *Nepenthes* pitcher plants are able to capture their prey – conveniently exaggerating their ability in his description:

The pitcher plant, for instance has a heavy flesh leaf ten inches long. With the spiked point of the leaf it strikes a rat, numbing it with the poison it contains. Then the leaf folds over the animal and it is absorbed into the body of the plant and digested . . . The pitcher plant devouring a rat, is an instance of plant life possessing animal powers.[38]

After providing this (highly exaggerated) 'proof' that Earth-bound carnivorous plants possess animal-like intelligence, Campbell goes on to make the extraordinary claim that Martian plants undoubtedly possess far greater levels of intelligence than Earth plants. (In doing so, of course, he totally disregards the fact that the Martian climate and atmosphere would be quite inhospitable to large plant life.) Much more common were articles that would highlight the quirky nature of the pitcher plants, including their greedy appetite for flies. One newspaper article, headlined 'Carnivorous Plants Killed by Indigestion', claimed that,

> A pitcher plant may be fed so many flies that it will die of an acute indigestion. One of these plants under observation has been fed two flies a day and thrived upon them. Three flies fed to it on one occasion caused it to show signs of distress, and when it had recovered from this indigestion five flies killed it.[39]

Another story about the display of Venus flytraps at Kew Gardens noted:

> These plants occasionally have been so overfed with flies by those eager to watch the working of their traps that they apparently have lost their sensitiveness, and refused to show off until the equanimity of their digestion had been restored. 'Don't you feed them!' commands the stalwart guard at Kew. 'At the present time their traps are not working.'[40]

Countless other articles exaggerated how these plants 'stalked' their prey. One article about the Venus flytrap noted how 'it steals along close to the ground, literally extending its "jaws" for the capture of insects'.[41] Another newspaper article erroneously claimed that, 'The *Sarracenia*, the *Nepenthes*, and the *Cephalotus* have lids which shut down upon their victims.'[42]

Picturing Carnivorous Plants

Carnivorous plants have a rich history of being depicted in highly detailed botanical illustrations. However, unlike other plant groupings, there has tended to be an intriguing shift in the manner in which these plants have been represented over the decades. Prior to the discovery of their carnivorous nature, the emphasis was, as with most plants, upon their specific morphology, structure and patterning. However, once it was proven that these plants were carnivorous, quite often imagery of their insect prey would also be incorporated in the illustrations. Obviously the fact that these plants are carnivorous is one of their defining characteristics. However, it seems somewhat inconsistent when you consider the fact that zoological illustrations of the time would rarely have depicted the prey of that depicted animal. Frogs or salamanders, for example, would not normally be shown to be eating flies; similarly a lion would not be seen preying upon a gazelle. In addition to the inclusion of their insect prey, some illustrators seemed inclined to imbue their carnivorous plant images with a touch of implied animation. Some images even appeared to suggest that the plant was in the midst of an aggressive attack.

Sundews were well known since at least the Middle Ages; however, the first published illustrations of these plants can be found much later, in the 1554 herbal book *Cruyde Boeck*, by Rembertus Dodonaeus; while the first published illustration of a *Sarracenia* pitcher plant can be found in the book by Mathias de L'Obel, *Nova stirpium adversaria*, published in 1576. The species depicted is most likely *Sarracenia minor*. An early detailed illustration of *Nepenthese* (likely *N. mirabilis*) can be found in the book *Herbarium Amboinense* by Rumphius, published

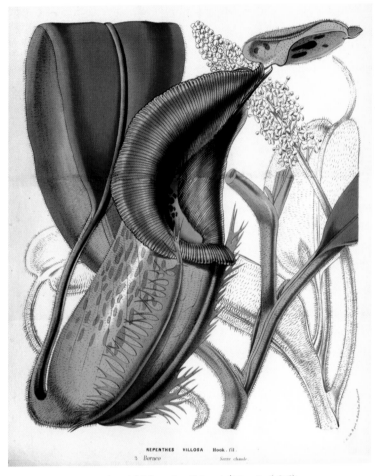

NEPENTHES   VILLOSA   Hook . fil .
3. *Borneo*   *Serre chaude.*

Nepenthes veitchii, from *Curtis's Botanical Magazine* (1858).

in 1747. It features detailed drawings not only of the pitchers, but of the plant's stems, roots and inflorescence.

James Roberts drew one of the earliest published, and highly detailed, images of the Venus flytrap. This illustration was also one of the first to show a carnivorous plant in action. The image depicted an earwig caught in one trap and a fly in another. It was published by John Ellis, in 1769, in his article 'A Botanical Description of the Dionaea Muscipula, or Venus's Fly-trap, A Newly-discovered Sensitive Plant.' At the time of its publication, it was well known that the plant caught insects, but it would not be proven to be carnivorous for another hundred years.

One of the most accomplished and prolific illustrators of the unique volumetric forms of the *Nepenthes* pitcher plants was the Belgian artist Pieter De Pannemaeker. Many of his illustrations were published in the French journal *Flore des serres et des jardins*. Horto Van Houtteano also produced a large quantity of highly detailed images of carnivorous plants for this same publication.

Curtis's Botanical Magazine was an important early publication that often featured full-colour illustrations of carnivorous plants – especially pitcher plants (*Nepenthes* and *Sarracenia*). In 1890 a *Nepenthes* plant was named after the founding editor, William Curtis, as the *Nepenthes curtisii*.

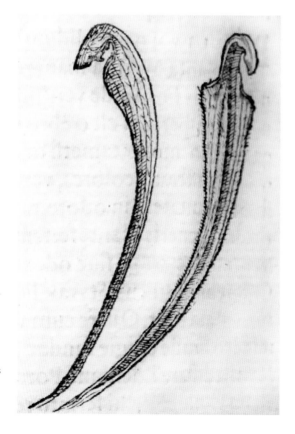

Believed to be the first published illustration of *Sarracenia*, in Mathias de L'Obel's *Nova stirpium adversaria* (1576).

Marianne North, *Nepenthes northiana*, c. 1880.

SARRACENIA DRUMMUNDI. Crom.
4 Floride 8 temp. S chaud.

Sarracenia illustration from *Flore des serres et des jardins de l'Europe* (1858).

The notable English artist Marianne North also had a *Nepenthes* species named after her. North was a highly prolific botanical illustrator and painter who would often travel the world, illustrating the native flora in habitat. While in Borneo, she painted a species of pitcher plant that had not yet been identified. The species was later named in her honour; *Nepenthes northiana*.

Carnivorous plants, from *Die Gartenlaube* (1875).

four
Attack of the Killer Plants
❦

Ever since their discovery, carnivorous plants have frequently
been depicted as violent, man-eating monsters. In reality,
most of these plants are minuscule – particularly those that
display any type of visible movement. It has taken a very substantial
leap of the imagination to aggrandize a small, insect-catching plant
into one that could devour a human.

But, of course, the human imagination is very good at embellish-
ment and exaggeration, and carnivorous plants have proven to be
excellent subject matter from which to work. In 1951 the editor of
the popular American sci-fi magazine *Startling Stories* noted that:

> Ever since the first of our arboreal ancestors studied and
> understood the function of a pitcher plant, or perhaps a
> Venus fly-trap, the idea of a vegetable growth which could
> and would entrap and absorb a human has been one of the
> well-springs of nightmare. For some reason the thought of
> becoming breakfast food for an outrageous orchid or perhaps
> an over bloated sweet potato holds more intrinsic horror than
> the fangs of the tiger or the tentacles of the kraken.[1]

There are numerous possible explanations as to why plants (particularly carnivorous ones) make such good subjects for horror stories.
One reason is that plants can appear to be fundamentally alien to
us; they have no central nervous system or brain and seem to move

A fly sits on the lid of a *Sarracenia* pitcher.

on a different timescale to our own. This can make it difficult for us to not only understand how they function, but identify with them.

Plants can also seem to be unnaturally persistent. In some cases we might find it nearly impossible to kill off a vine or a patch of weeds. Other plants that might appear to be dead (such as a withered vine or a leafless tree in winter) come roaring back to life like some sort of resurrected entity. Another potentially unsettling concept is that most plants seem to move only when we are not looking at them. Owing to the speed at which they develop, we never get to witness a plant actually growing; but returning to it after a few weeks, we might find that it has nearly doubled in size. In fact, some plants, such as the kudzu vine (*Pueraria lobata*), which has become an invasive weed in the southeastern United States, can grow nearly 60 cm (2 ft) in a single day, the vines capable of enveloping an entire house in a matter of weeks. This apparent ability to move or 'attack' when we have our backs turned effectively taps into one of our most cogent anxieties.

We also tend to think of plants as lacking sentience, yet carnivorous plants are able to kill and eat conscious living things. When it was first proven that these plants were carnivorous, some found the concept particularly disquieting. Writing in the 1880s, Colin Clout described how terrifying it was that in the battle between a sentient insect and an insentient plant, 'the insentient always wins' and that the 'breathing, flying, conscious insect' is slaughtered by 'cunning and apparent hunger for blood on the part of a rooted, inanimate plant'. Also, as noted earlier, the Italian criminologist Cesare Lombroso declared in his 1884 publication *Criminal Man* (third edition), that carnivorous plants are likely to be the evolutionary source of human 'evil' and that the human's propensity for criminal activity stems specifically from these plants.[2] With just a bit of a stretch of the imagination, one could consider that just about all plants are potentially 'man-eaters' in that they can benefit nutritionally from decomposing flesh. As one writer has noted, 'these man-eating plants only hyperbolize a mundane fact about our relationship with plants; however, in the end, we become their nourishment.'[3]

Early Carnivorous Tales

As part of their narrative strategy, most early fiction on carnivorous plants would purposefully guide the observer along an apparently logical path of amplification. First, the reader will be reminded of the many amazing real-life carnivorous plants, such as the Venus flytrap and its remarkable ability not only to eat insects, but perceptibly move. Then it will be suggested that it is quite feasible that there could be larger and more mobile versions of these plants still to be discovered in far-off lands. Or, alternatively, that a grossly transformed version of a Venus flytrap might potentially come about either as a result of swift changes in the natural environment, or through the intervention of some eccentric botanist.

One of the earliest descriptions of carnivorous monster-plants came from Dr Carl Liche, a German explorer who, in 1887, claimed to have discovered a man-eating tree in Madagascar. One newspaper described the tree as being:

> from seven to ten feet high and something like a grapefruit in shape with rough, ugly tendrils stretching out in all directions. The trunk is black and as hard as stone. At the top of the tree are palpi, six feet high, that rear straight up and twine and twirl about incessantly. There is a cup also at the top which contains a clear, appetizing looking fluid. But alas for him who drinks it. He becomes peculiarly crazed and unable to get down. Then it is that the whirling palpi twine themselves slowly but surely about the helpless man until all life is gone. This species of tree is naturally avoided as a deadly serpent would be, and the natives consider that it is actually alive and possessed of an evil and terrible spirit.[4]

Such newspaper reports, which seemed to be widely believed, were frequently published and republished over several decades and in a number of countries (America, Australia, England), each time captivating a new readership. Even as late as 1920 a U.S. newspaper article written by Dr B. H. William, 'The Distinguished American Botanist', detailed the likelihood of the existence of this tree. In concluding the article he writes: 'Intense as the interest the story of the Madagascar tree has aroused among botanists, there are few who will dare say it is impossible. It is to be hoped that a scientific expedition will soon go to the place indicated by the German explorer and make a careful study of the extraordinary plant.'[5]

Illustration of a killer plant from the *Ogden Standard-examiner, c.* 1900.

However, it was Sir Arthur Conan Doyle (better known as the author of *Sherlock Holmes*), whose classic short story 'The American's Tale' first helped to define and to popularize the 'carnivorous-plants-as-monsters' fiction genre. Published in December 1880 in the *London Society* magazine, the story describes how, on one cold rainy night, a strange American man named Jefferson Adams entered a London gentlemen's club and recounted a most astonishing story about giant Venus flytraps. These carnivorous plants were claimed to be so large that they could easily consume a person. He began his tale by enquiring,

> Now which of you gentlemen has ever been in Arizona? None, I'll warrant. I've been there, sirs, lived there for years; and when I think of what I've seen there, why, I can scarce get myself to believe it now. Maybe some of you has seen a plant as they calls the 'fly-catcher', in some parts of the States?[6]

To this, Dawson, the so-called scientist of the group, nods his head and murmurs knowingly the plant's Latin name, '*Dionaea muscipula*'.

'The Man-eating Tree of Madagascar', from J. W. Buel, *Sea and Land* (1887).

After this initial touch of scientific validation, the American continues, 'Well, I've seen those flytraps in Arizona with leaves eight and ten feet long, and thorns or teeth a foot or more.' He then proceeds to recount the story of how a man named Alabama Joe had gone missing one night and the next morning his remains were found, he had been devoured by a giant Venus flytrap. The scene is then described in graphic detail:

> One of the great leaves of the flytrap, that had been shut and touchin' the ground as it lay, was slowly rolling back upon its hinges. There, lying like a child in its cradle, was Alabama Joe in the hollow of the leaf. The great thorns had been slowly driven through his heart as it shut upon him. We could see as he'd tried to cut his way out, for there was slit in the thick fleshy leaf, an' his bowie was in his hand; but it had smothered him first. He'd lain down on it likely to keep the damp off while he were awaitin' for Scott, and it had closed on him as you've seen your little hothouse ones do on a fly; an' there he were as we found him, torn and crushed into pulp by the great jagged teeth of the man-eatin' plant.[7]

At the conclusion of the short story, Dawson again responds, affirming the plausibility of this tale: 'A most extraordinary narrative! Who would have thought a *Dionaea* had such power!' Significantly, this early monster-plant narrative by Conan Doyle laid the framework that most other monster-plant stories would follow. By initially introducing a touch of scientific plausibility, the audience would be gently primed to accept the possibility that more monstrous versions of these plants could exist.

Another influential short story at this time was 'Professor Jonkin's Cannibal Plant', by Howard R. Garis, which was published in 1905 in the British magazine *Argosy*. It tells of a botany professor who procures a unique specimen of pitcher plant from Brazil, which he refers to as a 'Sarracenia Nepenthis'. One of the exceptional features of this plant is that it has long hairs that are able to reach out, grab an insect prey, drop it into its pitcher, then snap shut its lid. The professor's assistant is totally amazed by this newly acquired plant – but the professor quickly becomes very secretive and bars the assistant from entering the greenhouse where the plant is kept.

Some weeks later, the assistant journeys to the market to stock up on provisions for the professor. He is astounded to learn from the grocer that the professor had been secretly ordering three large 'porterhouse steaks' on a daily basis, having them delivered to his greenhouse. 'But the professor and I are both vegetarians!' exclaims the assistant in disbelief. After being assured that it is indeed true, he muses aloud, 'Three porterhouse steaks a day! I do hope the professor has not ceased to be a vegetarian.' Rushing back to the lab, the assistant confronts the professor. To his relief, the professor assures him that he is still very much a vegetarian; then shows him the now gigantic 7-metre-tall (23 ft) carnivorous 'Sarracenia Nepenthis'. He explains:

> I reasoned that if a small blossom of the plant would thrive on a few insects, by giving it larger meals I might get a bigger plant... First I cut off all but the blossom, so that the strength of the plant would nourish that alone. Then I made out a bill of fare. I began feeding it on chopped beef. The plant took to it like a puppy. It seemed to beg for more. From chopped meat I went to small pieces, cut up. I could fairly see the blossom increase in size. From that I went to choice mutton chops, and, after a week of them, with the plant becoming more gigantic all the while, I increased its meals to a porter-house steak a day ... And now my pitcher plant takes three big beefsteaks every day – one for breakfast, one for dinner, and one for supper.[8]

A few days later, while feeding it an extra-large porterhouse steak, the professor is violently drawn into the gaping mouth of the plant. Fortunately his assistant soon hears his cries and commences to chop

Illustration from Howard R. Garis's 1905 short story 'Professor Jonkin's Cannibal Plant' from the magazine *Argosy*.

open the plant. But the professor calls out from inside the plant, pleading with him not to kill his 'Sarracenia Nepenthis'. Instead, the assistant douses the pitcher with chloroform gas; the plant falls limp and the professor is able to climb out. Relatively unscathed and only mildly annoyed, the professor then chides the plant. 'To punish it, I will not give it any supper or breakfast. That's what it gets for being naughty.' Of course, the reader is left with a rather uneasy feeling: a much more terrible horror is likely to occur – particularly if the vegetarian professor continues to deny the carnivorous plant its porterhouse steaks.

In the 1922 short story 'Drosera Cannibalis' by René Morot, the carnivorous plant narrative takes a decidedly macabre turn. The story begins innocently enough as it recounts numerous facts about carnivorous plants and describes the work of the main character, Dr Hartenstatter, who had

> made all kinds of observations on the growth and movement of the genus *Drosera,* plants which, as everyone knows, have the ability to capture the flies and other insects that alight on their leaves, closing over the victims immediately and in a few hours, thanks to the secretion of a very active pepsin, absorbing them wholly without leaving the slightest trace.[9]

Soon, Dr Hartenstatter develops a gigantic version, the *Drosera gigantis,* that grows as big as a tree. As the plant gets larger, so does its appetite, preferring first grasshoppers, then guinea pigs, then mice, then rabbits and finally lambs. But the plant soon craves more than just animal meat.

> One day Hartenstatter, deciding to give a new turn to his researches, brought back from the military hospital, where he went each morning to serve as a surgeon, the hand of a wounded man which had been amputated and which he had cleverly extracted from a heap of surgical debris destined for

Cover of magazine featuring the short story 'The Beast Plants' (April 1940).

destruction . . . his *drosera gigantis* absorbed the hand completely in five hours and forty-eight minutes, and although three times the same weight of food from a rabbit the week before had increased the trunk of the *drosera* only half a millimetre, the human flesh in this experiment increased it about two thirds of a millimetre. Dr Hartenstatter was triumphant. He had discovered a man-eating plant, *drosera cannibalis!*[10]

Some six months later, however, things become more dreadful when Dr Hartenstatter begins to abduct the village children and feed them to the plant. And, to justify his actions, the mad botanist declares 'if the vegetable kingdom were to disappear suddenly, the animals would die a few days afterward. The true life of nature is in the vegetable kingdom!' Finally the doctor is also consumed by the plant; the townspeople then burst into the greenhouse and burn down the monster and the lifeless remains of Dr Hartenstatter.

In 1930 H. Thompson Rich wrote the short story 'The Beast Plants', which was subsequently re-published in 1940 in the widely distributed American magazine *Famous Fantastic Mysteries*. The story recounts a journey into the swamps of southeast Georgia by a professor and his twenty-something daughter, Doris Mortimer, in order to continue his research work into the development of new species of carnivorous plant. The professor has a colleague, Neil Huntley, who some months later decides to travel down to visit him. Upon venturing into the swamps, Neil comes upon several giant Venus flytraps measuring over 6 metres (20 ft) tall. The plants quickly rush towards him and he is captured in their enormous traps. These mobile plants then carry him through the swamps and bring him to their queen, who, it turns out, is Doris. She quickly commands the plants to release Neil and welcomes him to her realm. She then describes how her father, who had since died, had developed these mutated *Dionaea*, which, in turn, had made her their queen.

In keeping to the conventions of the carnivorous plants horror genre, the two soon have an in-depth discussion on how the giant flytraps could plausibly have evolved as a result of her father's work. Neil speculates:

'Because it already had marked mobility of the fly-trap portion of its leaves, I can understand how he might have stimulated this mobility and also increased the plant's size. But I can't understand,' and Huntley was emphatic about this, 'how your father could have created an intelligence that didn't exist before.'

'But it did exist,' said Doris, almost with a shudder. 'Don't mobility and intelligence go hand in hand? Can anything move that hasn't nervous energy, and isn't that just another name for brain power? . . . Darwin hinted at it when he said that the impulse to motion of Dionaea is transmitted along the fibrovascular bundles. And currents flow through leaves much as they do through the nerves of animals.'[11]

Doris further explains how her father had carefully selected the hardiest specimens and fed them a course of chemical compounds that stimulated the growth of the leaves but atrophied the roots. This enabled the plants to gain mobility. He harvested the seeds from these newly ambulatory plants and soaked them in a solution to increase their size: '"Soaked them till they got as big as marbles," said Doris: "as big as baseballs!"' Finally, in order to increase their intelligence, he grafted in 'brain cells'.

Unfortunately the giant Venus flytraps begin to turn on the couple, but with a bit of luck and ingenuity they are able to fight them off. Eventually, the mutant flytraps are eradicated and the couple are able to escape.

The Day of the Triffids

One of the most famous stories about carnivorous plants is John Wyndham's novel *The Day of the Triffids*, first published in 1951. Set in England, this chilling book describes how the country is quickly overtaken by roaming swarms of man-eating plants known as triffids. The result of secret botanical experiments conducted deep within the Soviet Union, these mutant plants average some 2 metres (6½ ft) in height and are able to walk and communicate by making loud clattering sounds. They also possess a poisonous whip that can kill a large animal or human, after which they feed off the dead body. At the base of the plant are three root-like forms (thus the name, 'tri-fids') that can function either as traditional roots or be pulled out of the ground to serve as primitive legs. Also at their base are two small woody appendages known as clatter sticks which knock against the trunk of the plant. They use this knocking or clattering as a kind of Morse code by which to communicate with each other over long distances.

Movie poster for *The Day of the Triffids* (1962).

Still from the BBC TV series *The Day of the Triffids* (1981).

Importantly, the triffids produce large quantities of a very valuable oil, for which they are heavily farmed. Although the triffid's stinger can be docked to make it less dangerous, it was found that their oil production would then be greatly reduced. As a result, this practice was quickly abandoned and millions of deadly triffids were kept as plant-stock all over the world. Despite being 'blind', giving the human farmers a distinct advantage over them, the triffids would routinely try to attack the farmers with their stingers – always seeming to aim for the eyes (perhaps instinctively knowing that this was the source of the human's domination over them).

Then a great levelling event occurs. A strange worldwide meteor shower of exceedingly bright lights manages to blind nearly the entire world's population. The triffids suddenly have a distinct advantage over the sightless humans; they quickly begin to spread across the world, feeding upon mankind. A small number of people, including the main characters, Bill Masen and his companion Josella Playton, had managed to avoid being blinded and much of the novel is about their struggle to survive in a dystopian society and their efforts to help as many of the blind as possible. In the end a small group of people, both sighted and blind, make their way to the Isle of Wight and set up a colony there, safe from the advancing triffids.

Although the book goes into great detail about these killer plants, it also asks us to consider just what it is to be human, and to consider how easily this thin veneer of humanity can be eroded away. The triffids, although plants, are portrayed very much like animals and, as society crumbles, the humans also begin to act like animals. Ultimately *The Day of the Triffids* exposes how surprisingly fluid the boundaries between plants, animals and humans can seem.

Although focusing on the invasion of carnivorous plants, *The Day of the Triffids* also describes a compelling scenario in which plant life in general is able to assert its own version of evolutionary dominance. After a few years have passed, the main characters, Josella and Bill, decide to journey around the countryside. Many of the towns have become almost unrecognizable, having become heavily overgrown with plants and trees. In response to this resurgence of plant life, Josella remarks, 'It rather frightens me. It's as if everything were breaking out, rejoicing that we're finished, and that they're free to go their own way.' In this manner, one essayist has described the book as a sort of 'Darwinian parable' – one that suggests that it is arrogant to assume that humans are the pinnacle of evolution, and that we are unlikely to 'dominate perpetually'.[12]

From the book, three different film adaptations have been produced. The first movie version of *The Day of the Triffids* (directed by Steve Sekely) was released in 1962. This particular film strayed quite a bit from the original novel and also provided a decidedly upbeat conclusion. In this version, the surviving humans discover that salt water effectively kills the triffids, and thus huge quantities of ocean water are pumped inland to quickly destroy the entire population of the carnivorous plants. In 1981 the BBC televised a far more faithful representation of the book, a six-part miniseries (directed by Ken Hannam). Interestingly, the triffids in this production seem to have been modelled after the American pitcher plants (*Sarracenia*), but with added poisonous snapping whips that come out of their pitcher bodies. There was also a 2009 BBC production (directed by Nick Copus) which brought in many contemporary environmental concerns, highlighting, for example, how the triffid oil has become an almost total replacement for fossil fuels, virtually saving the planet from global warming. But in this version the triffids are also particularly deadly, possessing long vines that can stretch out at enormous speeds to capture their prey from great distances.

In 2001 a sequel to the original book was published, *The Night of the Triffids*, written by Simon Clark. In this novel, another strange worldwide event occurs – this time, the Earth is shrouded in darkness, covered by an enormous cloud. This veil completely blocks out the Sun, Moon and stars, effectively making even sighted humans blind. The triffids seem to take advantage of this very quickly and begin to invade the Isle of Wight. But the cloud of darkness soon lifts and much of this sequel then takes place in New York, where an American colony of resilient survivors has been set up to battle the triffids.

Little Shop of Horrors

The Little Shop of Horrors (1960) is a Hollywood B-movie, written by Charles B. Griffith and directed by Roger Corman. It describes how a strange and highly sentient carnivorous plant named Audrey Jr comes to live in a small flower shop on Skid Row. Famously, this low-budget film was shot in just two and a half days – two days on a borrowed set plus an additional night-time exterior shoot. Corman reminisced, 'I used to make films almost for the fun of it, as a joke. *Little Shop of Horrors* was essentially a gag to see if I could do it in two days and a night – and I did.'[13] This 'gag' was only moderately successful at first, but it went on to garner a large cult following. The film was adapted as an off-Broadway musical in 1982 and this, in turn, was re-adapted as a feature film in 1986.

In the original 1960 film version, the starring plant was the result of shop assistant Seymour Kelborn's careful hybridization work. It was described as a cross between a Venus flytrap and a 'butterwort'.

Having developed it at home, Seymour convinces his boss, Mr Mushnick, to allow him to display his strange new plant in the shop, suggesting that it might attract customers. The owner, though sceptical, finally agrees and Seymour places the plant prominently in the shop window, naming the plant Audrey Jr, after his co-worker, with whom he is in love. However, the plant begins to wither and look generally unhealthy, and its 'mouth' droops open at night. On one occasion, while attempting to care for his plant, Seymour accidently cuts his finger; drops of his blood fall into the plants open mouth. Immediately the plant begins to move, swallowing these drops of vital fluid. Realizing that human blood is what the plant needs, Seymour proceeds to cut all of his fingers and squeezes the drops into the hungry plants mouth. The next morning when he arrives, the plant has miraculously doubled in size. It continues to grow, as do the shop's fortunes as more and more customers visit the shop to see the unusual plant.

But the plant soon begins to wither again. One night, as Seymour studies an assortment of gardening books to see what might be the plant's problem, it calls out, 'Feed me!' Once Seymour gets over his initial shock, he holds up his bandaged fingers and says to the plant, 'Gee Junior, I'd like to feed you, but I've used up all my fingers!' But the plant becomes more insistent, demanding again, 'Feed me!' Not knowing what to do, Seymour goes for a late-night walk down by the railway yard. There, owing to his clumsiness, he accidently causes a security guard to stumble in front of an oncoming train. In a panic, he stuffs the severed remains into a bag and brings it back to the flower shop. There he finds Audrey Jr, even more insistent: 'Feed me! I'm hungry!' Seymour reluctantly feeds a severed hand to the plant, followed by a severed foot, which the plant hungrily devours. The plant is enormous the next day and attracts even more customers. Seymour continues to feed the plant human flesh – and soon comes under the plant's hypnotic control.

Meanwhile the local plant society plans to award Seymour its crowning prize for his spectacular plant. The ceremony is scheduled for the following evening – when the plant is likely to open its flower buds. The evening arrives; the flower buds open one by one – and it is revealed that each one contains the face of a human victim that the plant had devoured. Later, in an act of desperation, Seymour tries to kill the plant, but instead is eaten by it. In the final scene, a newly opened flower bud prominently displays the horrified face of Seymour Kelborn.

This original 1960 version of *The Little Shop of Horrors* very probably drew inspiration from the numerous horror-themed comic book stories that were popular at the time. Many of these comic books featured carnivorous plants. In particular, the film appears to draw inspiration from the October/November 1954 issue of the popular American comic book series *The Monster of Frankenstein*, published by Prize Comics. In this issue Professor Frankenstein has developed a carnivorous plant (not too unlike Audrey Jr). He begins by feeding it insects, but then orders his monster assistant to gather up mice and rats to satisfy its growing appetite. Strangely, the plant takes on the physical characteristics of whatever life form it eats. Thus when it eats the pet cat, its floral buds assume the form of a cat's head. Finally, the mad professor abducts the neighbour's baby and attempts to feed the infant to the plant, but he is stopped at the last moment by the Frankenstein monster. In a panic, the professor grabs the plant, straps it to his back and runs away. But the plant latches onto him as he flees, devouring him before he can free himself. The final panel shows the plant growing in the ground with numerous newly sprouted flower-buds, all in the shape of the professor's head.[14]

When the original Roger Corman movie was adapted as a Broadway stage musical, *Little Shop of Horrors*, opening in 1982, it featured a much more animated and verbose carnivorous plant. It also featured a more dramatic ending, with the plant devouring both Seymour and Audrey.

Owing to the show's success, *Little Shop of Horrors* was again adapted as a movie, directed by Frank Oz and released in cinemas in 1986. In this version, the carnivorous plant (this time named Audrey II) is

Seymour and Audrey II in *Little Shop of Horrors* (dir. Frank Oz, 1986).

portrayed in a far more anthropomorphic manner, highly sentient, able to speak fluently and also very dexterous and able to use its array of vines and leaves like hands (the production required over sixty people to operate a complex, oversized puppet). Interestingly, the highly anthropomorphic treatment of this particular Audrey character prompted one writer to note that 'as these plants become more human-like, they become not only about *plants* eating humans but also about cannibals eating humans.'[15]

Unlike the first film, where the plant was the result of Seymour's hybridization skills, this plant purports to come from outer space. Having had no role in its creation, Seymour understands very quickly that this creature is much more than a mere plant and is soon compelled to do its bidding. He makes a deal with the plant-monster by which if he provides it fresh human flesh, it will make him rich and famous. For a brief period, Seymour (and the plant) enjoy universal fame.

Despite all the fame and glory, Seymour realizes that to murder and to encourage the carnivorous nature of a plant is wrong. He soon dreams of getting away from it all, marrying his co-worker Audrey and living happily in the suburbs. He immediately asks Audrey to marry him, and she enthusiastically agrees. But the plant, attempting to put a stop to this, then tries to eat Audrey; Seymour rescues her just in time and the two escape.

But what happens next is potentially the most horrifying revelation of the film (and one that humorously references the widespread availability of Venus flytrap plants in the real world). As they run away from the shop, they are stopped by a mysterious man – 'Patrick Martin, Licensing and Marketing of World Botanical Enterprises' – who proposes to purchase the exclusive rights to 'take leaf cuttings and develop little Audrey II's and sell them in florist shops across the nation'. The man exclaims, 'Why pretty soon, every household in America could have one!' Audrey and Seymour are horror-struck. They refuse to engage in any part of the deal and chase the man away. Seymour then resolves to stop the plant, crying out 'Wait for me Audrey, it's between me and the vegetable.' He confronts the plant, accusing it of plotting world domination: '"Every household in America! Thousands of you eating, that's what you had in mind all along, isn't it? We're not talking about one hungry plant here; we're talking about world conquest!"' Greatly angered, the plant tries to eat Seymour, who, at the last minute, manages to electrocute it. Once rid of the plant, he and Audrey move to the suburbs to lead a quiet and peaceful life. A sinister final shot reveals a small Audrey II plant growing contentedly in their garden. As a final gesture, the plant turns towards the camera, and grins slyly.

Originally, this version of the film had a much grislier ending – and one that more overtly references the then current, real-world, Venus flytrap fad. However, it was re-edited after performing poorly in pre-release screenings. In the original ending (which was made available on the DVD release), Seymour and Audrey are both eaten by the plant. Following the death of these main characters, the Audrey II plants become a worldwide phenomenon. People flock to the shops across the world to buy the adorable seedlings and soon nearly every home has one. But suddenly these cute little seedlings mature into monster-plants and they rapidly grow to nearly ten storeys tall. Quickly they take over the world, destroying entire cities like a multitude of

Godzilla monsters. In a final spectacle, a giant Audrey II appears to burst through the cinema screen and into the theatre: the threat of these carnivorous plants is thus, momentarily, made magnificently real to the viewing audience.

After the box-office success of *Little Shop of Horrors*, rather than heeding the film's parodic warnings, sales of Venus flytraps increased exponentially as the general public rushed to their local nurseries to buy their very own little 'Audrey' plant.

There was also a short-lived animated television series adaptation of the 1986 movie, simply named *Little Shop* (1991). This cartoon series was pitched towards a decidedly younger audience, with Seymour also depicted as a young teenager. He works part-time at Mr Mushnik's flower shop and pines after the equally young Audrey, who is Mr Mushnik's daughter. The plant, simply known as Junior, was sprouted from a 200-million-year-old seed which Seymour had found deep within an underground cavern. As the plant grows, Seymour discovers that it can talk and move – but he keeps this a secret. Junior happily takes up residence at the flower shop and, although it is full of attitude, the plant is kind to Seymour, helping him with his many day-to-day problems. In this sanitized version, although Junior is a carnivorous plant, it would much rather eat a platter full of hamburgers than a human. Besides being able to talk and walk, the plant also has what is referred to as 'vegetable magnetism', allowing him to telepathically control anything derived from plant matter – such as wood, paper and cotton. The show is replete with plant references, puns and gags. In one episode, Seymour introduces Junior to a conventional Venus flytrap plant with which Junior falls immediately in love.

Body of the Prey (or Venus Flytrap)

The low-budget horror-movie *Body of the Prey* (dir. Norman Earl Thomson, written by Ed Wood, 1970, aka *Venus Flytrap*) features a mutant carnivorous plant monster. The film is rather clumsily designed and replete with poorly phrased dialogue – often seeming to be ad-libbed by the actors. The lead character is a NASA scientist, Dr Bragan, who is suffering from exhaustion and overwork. His Japanese colleague recommends he take a lengthy holiday to Japan, inviting him to stay with his female cousin, Noriko. Knowing that Dr Bragan has a great interest in botany, he is offered a secluded greenhouse and laboratory to work in – with Noriko serving as his assistant.

On his way to the airport he has car trouble. He stops over in a small town in North Carolina where he digs up a native Venus flytrap plant, which he then takes with him to Japan. Once established in his Japanese laboratory, he begins work on developing and transforming his Venus flytrap. Dr Bragan gives Noriko a detailed lecture on its workings. He concludes his oration proclaiming, 'Charles Darwin called it the most wonderful plant in the world.' Then, in an extraordinary leap of logic, he muses, 'Perhaps, he had another theory about the origin of life. If this plant can think . . . and reason . . . then why can't it be human?'

A few nights later, Dr Bragan speaks directly to the plant, 'You think, you feel. You must be, in part, human. But like all humans you are weak. But, I'll find a way. Mark my words, I'll find a way. You will become the most powerful thing *on this universe* [*sic*].' Then after an ominous flash of lightning, he proclaims, 'Your mother was the soil. Perhaps the lightning will become your father!'

Dr Bragan devises an elaborate plan to graft together the Venus flytrap with a plant that he refers to as the *Venus vesiciloso* (a fictional tubular plant that lives on the ocean floor and 'devours hapless fish and other marine life'). After careful grafting and a surge of electrical current from a lightning storm, the plant transforms into a decidedly humanoid creature with giant flytrap hands and feet. Bragan names it Insectivorous in a tribute to Charles Darwin, but, after a few days, the transformed plant becomes sickly; Bragan starts feeding it live rats, rabbits and chickens and it soon comes back to life and flourishes.

Not yet satisfied with his creation, Dr Bragan decides that in order to make it truly sentient it will require the 'blood of a human heart'.

'Pitcher Plant', from Robert John Thorton's *The Temple of Flora* (1807).

That night he sneaks into a nearby hospital and extracts some blood directly from the heart of a sleeping patient. As the plant-monster Insectivorous continues to grow, it also begins to crave human flesh. It is at this point that Noriko can no longer endure these horrifying events and she exclaims, 'This thing is a monster! You are no longer Dr Bragan, scientist. You are becoming Dr Bragan, mad-man!' When the plant-creature escapes from the laboratory and heads towards the village, Dr Bragan chases after it and, fortunately for humanity's

Illustration of various carnivorous plants in habitat, from Anton Joseph Kerner
von Maurilou and Adolf Hansen's *Pflanzenleben*, vol. 1 (1913).

sake, both he and the plant-monster fall from a mountainside and drown in a pit of flowing lava.

More Killer Carnivores

The popular 1950s television series *Alfred Hitchcock Presents* briefly featured a carnivorous plant. In one episode, dating from 1959, Alfred Hitchcock humorously (but calamitously) interacts with a giant carnivorous plant during the episode's opening and closing sequences. In the opening sequence, Hitchcock stands next to an enormous lily-type flower and, in his customarily droll delivery, addresses the camera, saying, 'Good evening ladies and gentlemen. I'm sending this beautiful plant to a dear friend.' He produces a large slab of raw meat – 'I believe it's feeding time' – depositing this into the flower's mouth, which proceeds noisily to gulp down the food. Hitchcock continues: 'These carnivorous plants get quite hungry. This one has been quite useful around here as a garbage disposal. I shall hate to part with it but I know my friend will love it, *and I am sure it will love him too!*'

At the close of the episode, Hitchcock is seen again standing next to his carnivorous plant, this time holding a bottle of perfume in his hand. Spraying the flower with the perfume he declares, 'I'm

An episode of *Alfred Hitchcock Presents* featuring a 'carnivorous flower'.

attempting to improve on nature by giving this flower a more inviting scent. I want my friend to get quite close!' As Hitchcock continues spraying the perfume, he accidentally drops the whole bottle into the flower. In an attempt to retrieve it, he reaches his arm into the flower and is devoured by the plant.

The American television series *The Addams Family* featured a pet carnivorous plant known as an 'African strangler' named Cleopatra. There have been several versions of *The Addams Family* show: one aired in the 1960s, in 1992 there was an animated version and another live-action version was produced in 1998. This carnivorous plant is regularly fed bite-sized chunks of meat. In the original 1960s series, the plant does not look particularly anthropomorphic and the food is merely passed by fork through its leaves. However in the 1998 series, the plant looks very much like Audrey II from the *Little Shop of Horrors* movie (1986). In this latest version, Cleopatra has a ravenous appetite and loves snacking on fresh emu eyeballs, rats, table scraps and anything meaty that she can find. In one episode, 'The Green of the Nile', the plant is rather ironically put on a strict diet of water in preparation for a plant show. In the end, Cleopatra does win the contest, earning an award for 'Most Vicious of Show'. The animated television series from 1992 also features Cleopatra; in this iteration, the plant has heavily anthropomorphized features, including a large human-like mouth with large lips and a pink tongue.

The *Batman* comics and movie franchise feature a supervillain known as Poison Ivy. She is a crazed botanist who has a decidedly ecological and pro-botanical agenda. Poison Ivy is happy to attack any human (including Batman) if she thinks that they are in some way harming her precious botanical beings. She is also able to use mind-controlling botanical pheromones against humans; in some cases, she can communicate with plants, using them to help her when battling her enemies. Being a highly talented scientist, she has also developed a number of mutated species of carnivorous plants – sometimes cross-breeding them with highly poisonous plants, or even with predatory mammals.

Similar to *Batman*'s Poison Ivy is the character known as Venus McFlytrap in the *Monster High* franchise. This fictional world comprises a lengthy series of animated movies and an extensive line of toy dolls. All the *Monster High* characters are ghouls and monsters, Venus McFlytrap being described as a zombie-creature, the 'daughter of the Plant Monster'. Venus McFlytrap can also control plant life (she can make large vines grow and creep along at such speed that they can overtake and smother an enemy). Furthermore, she can exhale clouds of pollen onto other monsters in order to control their minds. Similar to Poison Ivy, she is overt in her ecological agendas, often wearing badges or T-shirts inscribed with such phrases as 'eco-punk'. She, like many of the characters from the *Monster High* universe, has an associated pet. Hers is a Venus flytrap-like character named Chewlian, which has large teeth and can be quite vicious.

The *Super Mario Bros* video-game series and associated animated television series feature Venus flytrap-like carnivorous plant creatures. These are known as piranha plants. Most of these creatures reside in pots and have limited mobility, although a few, such as Petey Piranha and Dino Piranha, which have animal bodies with flytrap-styled heads, are much more animal-like. Piranha plants were first introduced as minor obstacles to overcome in the video game *Super Mario Bros.* (1985), and later took on more complex roles in subsequent Mario Bros games. In the 1990s animated television series based on the video game, they often appear in large numbers. In one episode, the character Luigi exclaims to Mario, 'I've heard of weed-killer . . . but *killer weeds?!*' as swarms of the plants advance across a field.

The *Plants vs Zombies* video-game series involves ongoing battles between human zombies and mobile plants. One of the plants featured is Chomper, a Venus flytrap-inspired character. In this pro-plant game-world, Venus flytraps are considered to be virtuous characters that can be employed to help rid the world of evil human-zombie characters. Chomper is capable of capturing and eating an enemy zombie in a single gulp, usually leaving one dangling arm visible from its mouth as it chews. Similar to the real flytrap plants, it takes Chomper

Venus flytrap consuming a captured insect.

time to digest its prey before it is able to 'chomp' again. In the sequel, *Plants vs Zombies: Garden Warfare*, the Chomper character is highly mobile, travelling to various locations to seek out its zombie prey.

There are several carnivorous plant-themed characters in the popular Japanese animated series *Pokémon*, too. One, Carnivine, is based on the Venus flytrap plant. Called a plant-creature, it is known as a 'bug catcher Pokémon'. It lives in swampy areas and 'gives off a sweet aroma that lures others close . . . then it attacks'.[16] Another character, known as Victreebel, is based on the tropical pitcher plant *Nepenthes*. Victreebel 'uses its long vine like a fishing lure, swishing and flicking it to draw prey closer to its gaping mouth'. These creatures again have plant-inspired weaponry: the 'razor leaf attack', the 'leaf tornado' and the 'vine whip'.[17] Similarly the *Digimon* franchise includes such characters as Blossomon, once again based on a Venus flytrap. Blossomon, in addition to its deadly flytraps, possesses razor-sharp leaves which it can fling at its opponents.

There are countless additional examples of killer carnivorous plants that can be found throughout popular culture. While some of these tales are presented, and intended to be received, in a very serious manner; many others are imbued with a healthy dose of humour and irreverence. Yet regardless of the form that these stories take, they will inevitably make special reference to real-world carnivorous plants – making these fantastical killer-carnivores seem, at least on a conceptual level, feasible.

The lower pitcher of *Nepenthes northiana*, native to Borneo, was named after the English artist Marianne North.

five

Magnificent Carnivores

ᕙ❀ᕗ

D espite their rather troubling reputation, carnivorous plants are truly magnificent configurations. Admittedly, a few species might be characterized as having slightly menacing characteristics: Venus flytraps display some rather spiky looking 'teeth'; the pitchers of *Nepenthes bicalcarata* develop what look like sharp fangs; the pitchers of *Darlingtonia* can resemble the highly venomous cobra, poised and ready to strike. Yet when we separate these plants from their mythological killer reputation it is easy to accept how truly stunning they are. One would be hard-pressed to find anything as beautiful as the brightly coloured pitchers of *Sarracenia leucophylla*, or the glistening and colourful dewdrops of a *Byblis* (rainbow plant) as it reflects and refracts the sunlight. Then there are the amazing colorations and ingenious mechanics of the Venus flytrap, the remarkable formations of the *Nepenthes* pitchers which dangle from quite unassuming vines, or the delicate beauty of the floral displays of *Pinguicula* (butterworts) and *Utricularia* (bladderworts)

Although such favourable portrayals have typically been avoided in mainstream culture, carnivorous plants have been greatly admired in traditional cultures and are being increasingly celebrated in art and design. These beautiful plants are also being used increasingly as home and garden features, and even in cut-flower arrangements. They have played a role also in both traditional and contemporary food and medicine. Remarkably, these extraordinary plants have even provided inspiration to industrial engineers.

Contemporary Carnivorous Art

Some of the most evocative portrayals of carnivorous plants are to be found in the paintings of Madeline von Foerster, a contemporary artist who employs traditional sixteenth-century Flemish techniques of oil and egg tempera to produce meticulously detailed images. Many of her paintings showcase beautiful, yet sometimes unsettling, images of fauna and flora – including carnivorous plants. These often engage with contemporary environmental concerns of deforestation, endangered species and loss of habitat. They also seek to address how humans have frequently fetishized certain aspects of the natural world while simultaneously destroying the broader environment. 'We love nature, we kill nature, and we can't quite figure out our relationship with it,' sums up von Foerster.[1] She describes her artworks as 'living still-lifes, which intentionally use the motifs of that genre to explore our assumptions about ownership and objectification of nature. But on a deeper level, they are visual altars for our imperilled natural world.'[2]

Three of her paintings which most prominently showcase carnivorous plants are *Carnival Insectivora (Cabinet for Cornell and Häckel)* (2013), *My Darlingtonia* (2009) and *Donne Unica* (2011). Of these paintings, von Foerster notes that each one 'speaks to my obsession and delight with these precious plants, as well as my desire to protect them from harm'.[3]

Carnival Insectivora (Cabinet for Cornell and Häckel) features a carefully contrived display of carnivorous plants. The painting depicts a wide range of species which are placed together in a shallow pot and presented within a simple whitewashed wooden display box. The portrayed plants comprise *Drosophyllum* (dewy pine), *Darlingtonia* (cobra plant), *Dionaea* (Venus flytrap), multiple species of *Sarracenia* (American pitcher plants), *Drosera* (sundew), *Heliamphora* (sun pitcher) and *Nepenthes* (tropical pitcher plant). In addition to these plants, two hands are shown gently embracing the potted display. There is a strange and surreal quality to these hands, which appear to be entirely dislocated from their human body. In fact, because of their position and their

Madeline von Foerster, *Carnival Insectivora (Cabinet for Cornell and Haeckel)*, 2013, oil and egg tempera on panel.

porcelain white sleeve-cuffs, they almost appear to be appendages of the ornate white porcelain plant pot. The embracing action of these hands is intended to convey 'a protective gesture, since so many of these living gems are also threatened by over-collection and habitat loss'.[4] However, because of our rather complex involvement with nature, von Foerster also offers an alternate reading:

> I suppose a darker interpretation of this gesture could be that the hands are 'hoarding' the plants, which is unfortunately fitting, since some species are in danger of being 'loved' to extinction . . . It is ironic that our adoration will be either the undoing or the saving of what's left of wild carnivorous plants.[5]

This spectacular display of carnivorous plants is further enhanced by the addition of several jewelled insect forms. A spider made of gold and precious gems hangs down from a pearl chain; a jewelled dragonfly rests on the side of a pitcher plant. Also, one of the hands is placing a beetle made from precious stones on to a *Nepenthes* pitcher plant. Although the addition of these exquisite insect forms enhances the botanical display and perhaps signifies an expression of human adoration, they would be of very little use to a carnivorous plant. Similarly, being contained within such a small box may make for a stunning display, but if kept in this manner the plants would have little hope of surviving, let alone prospering. These plants appear to be imprisoned, yet lovingly so.

The painting titled *My Darlingtonia* features an unclothed woman holding a *Darlingtonia* pitcher in her hand. With her other hand, she can be seen delicately placing a strand of pearls into the mouth of a *Sarracenia* pitcher plant. The woman wears a pearl earring and the *Darlingtonia* pitcher is similarly adorned with a pearl – which dangles from the plant's protruding 'tongue'. The painting depicts a rather impossible scenario, in which the relationships between the human, insects and carnivorous plants seem to provide more questions than

Madeline von Foerster, *My Darlingtonia*, 2009, oil and egg tempera on panel.

answers. Resting on a large plate in front of the woman is an enormous, dead fly. Yet rather than feeding this surreal insect to the plants, she appears to be feeding the pitcher plant a diet of pearls – a most expensive snack, which is, of course, of little nutritional value.

Donne Unica, or 'Unique Ladies', is another painting that features a wide assortment of carnivorous plants. In this instance they are merely leaf cuttings arranged in a floral-like display. The cuttings sit within a large Romanesque pitcher-vase that is emblazoned with a female figure and the text 'Donne Unica'. The carnivorous plants are rendered in exquisite and alluring detail, and a fly can be seen resting on the table beside the plant display. In this painting the artist 'wanted the viewer to feel as helplessly entranced as the fly' by the plants' 'dangerous allure'.[6] However, since the display consists only of leaf-cuttings, it is unlikely that the fly is in too much danger; and, as with cut flowers, the beauty of these carnivorous leaves will be

Madeline von Foerster, *Donne Unica*, 2011, oil and egg tempura on panel.

fleeting. Nevertheless, the painting represents a truly magnificent display of carnivorous plants.

Jane Ianniello is an Australian artist who imaginatively combines cinematic images from Hollywood film noir of the 1940s and '50s with vibrantly coloured and grandly scaled botanical images of carnivorous plants. Her paintings, which she refers to as 'noirscapes', are 'designed to unsettle the viewer' while evoking a sense of 'mystery'.[7] The disconcerting strength of these paintings resides within the incongruous juxtaposition of the two very different subject matters. Taken individually, each element is fairly innocuous. The carnivorous plant imagery, which includes colourful images of *Sarracenia* (American pitcher plants), *Drosera* (sundew) or *Dionaea* (Venus flytrap), is accurate in their botanical depiction. The appropriated cinematic component

Jane Ianniello, *Trapped*, 2016, acrylic on stretched canvas.

features classic Hollywood actresses imbued with, perhaps, a touch of dramatic melancholy, but which are otherwise faithful in their treatment. However, it is the exaggerated relative scale of the botanical forms next to the human form that inevitably delivers their most disturbing effect. This highly unbalanced juxtaposition simultaneously amplifies the beautiful features of the carnivorous plant as well as its predatory reputation.

Glass artist Evan Kolker has created a series of glass and metal sculptures that depict *Nepenthes bicalcarata* pitcher plants.[8] The sculptures are exquisitely crafted in glass with considerable botanical detail, comprising the lid, fangs, ridges and wings of the pitcher. The stems and the conventional photosynthetic portion of the leaf are constructed of sculpted steel, contrasting the pitchers made of blown glass and delicately suspended from the steel leaves by thin glass strands. Some of the sculpted pitchers are made from coloured glass, adhering more or less to the plant's natural colorations. Others are made from clear glass, allowing the viewer to peer into them. In a hyperbolic reference to their carnivorous nature, the clear glass pitchers also contain fragments of animal bones – suggesting the gruesome remains of the plants' recent carnivorous feasting. Each sculpture is also given a somewhat whimsical addition to their Latin name, which accentuates their carnivorous or lively qualities, such as N. *Bicalcarata in Cognito* and N. *Bicalcarata in Animus*.

Jason Gamrath is a glass artist from Seattle, Washington. Most of his glass sculptures depict oversized botanical forms, including numerous carnivorous plants. A great admirer of nature, by his dramatic amplification of botanical forms Gamrath seeks to direct our attention to the amazing and wondrous complexity, to the 'perfection' that is found within the plant kingdom.[9] His blown-glass pitcher plants (*Sarracenia*) measure well over a metre in height. Being made of glass, these sculptures transmit light and appear to illuminate from within in much the same way as the natural pitchers as they seek to attract insects to the plant. To make such forms out of blown glass, and on this scale, requires a great deal of skill. But, in addition to his

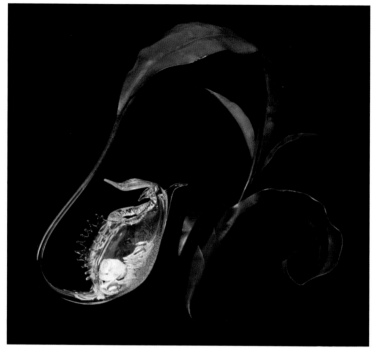

Evan Kolker, *Nepenthes Bicalcarata Stans Solus*, 2013, glass, steel.

technical expertise, the artist also manages to craft the plant forms with remarkable detail and sensitivity. He has also created a giant Venus flytrap with stems made of steel, the traps of blown glass. Gamrath's stunning sculptures have been exhibited widely in both traditional galleries and non-traditional settings. On several occasions, they have been exhibited in botanical gardens and displayed among the real carnivorous plants.

Using not glass, but a similarly translucent fibreglass material, artist Dan Corson has created a series of large *Nepenthes*-inspired outdoor sculptures. These, standing over 5 metres (16 ft) high, 'are inspired by the shape of a variety of *Nepenthes* and celebrate the wonderful diversity and quirkiness' of the local cityscape.[10] The sculptures are each nearly identical in form, but showcase a wide range of brightly coloured patterns. On display in various sites throughout Portland, Oregon, these sculptures also serve as light posts, the illumination of which further transforms in colour and pattern over time. The artist particularly enjoys working with light, 'creating environments that radically transform from day to night'.[11] During the day, the sculptures capture and transmit sunlight through the translucent fibreglass. Each sculpture contains solar cells and batteries which store electricity during the day, at night illuminating the forms from within. When lit up, the giant pitchers gradually change both in colour and intensity.

Paul Hill is a sculptor who lives and works in Venus flytrap country – North Carolina, USA – and has fabricated a series of sculptures

Jason B. Gamrath, *Sarracenia Pitcher Plants*, 2013, blown and sculpted glass.

Dan Corson, *Nepenthes*, 2013, fibreglass, paint and mixed materials.

Paul Hill, *Natural Embrace* (left), 2014, corten and fused glass; *Southern Hospitality* (right), 2014, carbon steel and fused glass.

that depict his native plant. The sculptures are enormous, standing nearly 5 metres tall, and are constructed primarily of steel but also incorporating fused glass.[12] One of these sculptures, rather whimsically named *Natural Embrace* (2014), has stems constructed of oxidized steel that have turned reddish in colour. By contrast, the traps are made of fused bright-green tinted glass. The stark difference between the rust-coloured stems and the green glass suggests the plant's more common coloration. Arranged in a very animated pose with the traps splayed out theatrically, the plant's open traps seem quite amenable to performing their 'natural embrace'. By physically magnifying a Venus flytrap form to such a large scale, these sculptures inevitably tap into the popular cultural mythology of man-eating plants. Their immensely large form makes it plausible, at least conceptually, that such plants could devour a human.

Another sculpture in Hill's Venus flytrap series, *Southern Hospitality* (2014), features intricately designed and polished steel stems – produced in an almost filigree pattern. These traps are also made of pressed glass that interplays with both the abundant sunlight during the day and the bright spotlights which shine upon them at night.

The wry title references the reputation that the local residents of the South are friendly and hospitable. But, of course, a Venus flytrap is never very hospitable to its insect guests – and such a giant flytrap as this would also be less than hospitable to any human guest who might stumble upon it.

Carnivorous Culture

The purple pitcher plant (*Sarracenia purpurea*) is the provincial flower of the Canadian province of Newfoundland and Labrador. This carnivorous plant (sometimes referred to as the 'side-saddle flower') is commonly found throughout the area – a region that also marks the northernmost latitude of the *Sarracenia* genus. Queen Victoria is credited with first suggesting that the plant become Newfoundland's official flower. Subsequently the 1865 one-cent coin became the world's first currency to feature a carnivorous plant, illustrated with a wreath of the plant's leaves and flowers on its reverse. Later, in 1938, the design was changed to feature a more conventional and detailed view of a *Sarracenia purpurea*. This newly designed coin was minted annually until 1947, when Newfoundland adopted Canadian currency. A stylized representation of the plant is now the official 'wordmark' logo of Newfoundland and Labrador, appearing on signs, government publications and even car registration plates.

More recently a number of Southeast Asian countries have also featured *Nepenthes* pitcher plants on their currencies. In Malaysia, the 20-ringgit note features tropical pitcher plants, as does the 5-dollar note from Brunei. Between 1998 and 2016, the 100-rupee note from the Seychelles also featured *Nepenthes*, and the current 5-rupee coin continues to do so.

Since the 1960s carnivorous plants have also been featured on numerous postage stamps from dozens of countries across the world. Many Asian countries, including Malaysia, Laos, the Seychelles, Singapore, India and Palau, have issued stamps featuring their indigenous *Nepenthes* species. Australian stamps have displayed their own native carnivorous plants with a set of four 60-cent stamps issued in 2013, each featuring a different species: *Nepenthes rowaniae*, *Cephalotus follicularis*, *Drosera lowriei* and *Drosera rupicola*. In addition to images of the plants, these stamps also included depictions of the plants' potential prey. Thus the *Nepenthes* was paired with frogs, the *Cephalotus* with ants and the *Drosera* with flies and butterflies. Australia also issued a 2-dollar stamp featuring *Drosera microphylla* (known as the 'golden rainbow' sundew). The United Nations produced a 32-cent stamp in 1996 depicting the Australian pitcher plant, *Cephalotus follicularis*, in its Endangered Species series.

The United States issued a series of four different 34-cent carnivorous plant stamps in 2001. Each of these stamps depicted a different carnivorous plant along with an insect that was either caught in the trap or about to be caught. On this series of stamps, only their common names were noted: yellow trumpet, Venus flytrap, English sundew and cobra lily. In 1966 Canada issued *Sarracenia* stamps, and later in 2006 it produced stamps featuring *Pinguicula*.

Many other countries have featured carnivorous plants on their postage stamps. While in some cases the featured species will be of native plants, in others they appear to be merely celebrations of the

One-cent coin from Newfoundland, 1941, featuring an image of *Sarracenia purpurea* on the reverse.

unique characteristics of carnivorous plants from all regions. Guyana issued a one-cent stamp in 1971 featuring its native sun pitcher (*Heliamphora nutans*) with the text, 'Pitcher plant of Mount Roraima'. In 1978 Ireland issued a stamp depicting *Pinguicula grandiflora*, and in 1995 stamps showing various *Sarracenia*. Saint Pierre and Miquelon issued a *Sarracenia*-themed stamp in 1962, and in 2007 produced stamps featuring *Sarracenia*, *Pinguicula*, *Drosera* and *Utricularia*. Japan issued a stamp featuring its native *Pinguicula ramosa* in 1978.

More recently, several countries have begun to produce postage stamps that survey a wider variety of the world's carnivorous plants – proving their increasing global popularity. In 2000 Somalia issued a series of stamps featuring *Drosera bulbosa*, *Drosophyllum* and *Dionaea muscipula*. St Vincent and the Grenadines issued stamps in 2005 that depicted *Nepenthes*, *Pinguicula*, *Dionaea muscipula* and *Drosera*. Guinea Bissau, in 2014, issued the following stamps: *Cephalotus follicularis*, *Sarracenia oreophila*, *Pinguicula gigantea*, *Brocchinia reducta* and *Heliamphora*. In 2015 São Tomé issued a series of stamps featuring *Nepenthes aristolochioides*, *Nepenthes rajah*, *Drosera roraimae*, *Heliamphora ionasi* and *Dionaea muscipula*.

In Malaysia, Singapore and other parts of Southeast Asia, the pitcher plant has become a celebrated icon. References to these plants are frequent and can manifest in surprising ways. Victoria, capital city of the Seychelles archipelago, is dotted with concrete, *Nepenthes*-shaped dustbins. In the Merdeka city square of Kuala Lumpur, Malaysia, there stands a large *Nepenthes* sculpture and fountain, known locally as the Periuk Kera Fountain. The fountain features eight giant *Nepenthes* pitchers which wind around a large tree trunk. The fountain's falling water fills up the top pitchers, which in turn fill the lower pitchers, then overflowing into the pool below.

Botanical gardens throughout the area will often showcase stunning displays of pitcher plants. In Singapore, the carnivorous plant garden is one of the more celebrated exhibits at the Gardens by the Bay. Here, large billboards have advertised the garden's membership programme and the many benefits that it affords – at the same time promoting its carnivorous plants display. One sign depicted a colourful

A large *Nepenthes* sculpture and fountain, known locally as the Periuk Kera Fountain.
Located in the Merdeka city square of Kuala Lumpur, Malaysia.

grouping of *Sarracenia* pitchers, accompanied by the text 'Drink more
for less'. Another sign displayed a large Venus flytrap, proclaiming
'Chomp down on our deals'. Similarly, tourist gift shops across the
Malay Peninsula sell *Nepenthes*-themed souvenirs, which can range from
pens and paperweights to clothing and homewares as well as genuine
pitcher cuttings preserved in blocks of transparent resin. For the

Advertising with a carnivorous-plant theme in Singapore.

more adventurous, a number of carnivorous plant adventure tours are available that guide small groups through the mountainous jungles of Borneo and other Southeast Asian localities.

Eating Carnivorous Plants

Although carnivorous plants have a mythical reputation for eating humans, there are some instances in which humans have been known to eat carnivorous plants. Various species have been used in traditional and homeopathic treatments and there is a long-standing tradition in Malaysia of using *Nepenthes* pitchers in cooking a glutinous rice dish.

The use of *Nepenthes* pitchers in cooking rice goes back centuries, but perhaps the first European explorer to document this was Frederick Burbidge, who wrote in 1897, 'When I was staying with the headman of the Kadyans on the Lawas river, his people often gave me delicious rice cooked in the pitchers of *N. hookeriana*, as a sweetmeat to be eaten with jungle fruit or bananas.'[13] Locally it is referred to as *pulut periuk kera* (pitcher plant rice). The pitchers themselves are generally not eaten, but simply used in the same manner as banana tree leaves in preparing Malaysian lemang and similar sticky rice dishes. It is claimed that the pitchers add a distinctive flavouring to the rice. Freshly picked pitchers are required in preparing *pulut periuk kera*. These are thoroughly washed out (removing any insect debris that might be inside); the tendrils, lids and 'wings' of the pitchers are removed, as is the peristome (the lip around the mouth opening). They are then filled with a mixture of rice and coconut milk and are ready to eat after steam-cooking for approximately forty minutes.

Although it is seen as a traditional and somewhat localized cuisine originating among a few indigenous groups, pitcher plant rice has gained wider popularity in recent years. It can often be bought in markets and from street vendors. In 2013 the dish gained a high level of notoriety when it was featured in Malaysia's very popular and long-running animated television series *Upin & Ipin*. In one episode, the

Rice-filled *Nepenthes* pitchers, known as *pulut periuk kera*.

main characters (twin boys named Upin and Ipin) learn about these pitcher plant rice treats: how they are made, their history and, importantly, how delicious they are.

Nepenthes pitcher plants have been widely used in traditional remedies. The fluid from unopened pitchers has been consumed as a handy source of drinking water when travelling through the local forests and has been employed to relieve a variety of ailments, including stomach problems, breathing problems and coughs. The stems and roots were sometimes cooked and used to treat malaria and stomach pains.[14] Chinese medicine derived a mixture from the plants of *Nepenthes mirabilis* as a treament for urinary tract infections and high blood pressure.[15]

More recently, several different varieties of carnivorous plant have gained popularity in homeopathic therapy (although it should be noted that, as of yet, none of these have been been scientifically proven). A widely available supplement is a Venus flytrap extract which is purported to be helpful as a supplementary treatment for certain kinds of cancer. One manufacturer of this supplement, which markets their product under the name 'Carnivora', claims that it 'inhibits tumor growth' and helps in 'stimulating immune response'.[16] While these claims are unsubstantiated to date, there has been at least one scientific study that has hinted at the possibility of the Venus flytrap having some beneficial anti-cancer properties. The study found that the plant contains a number of recognized anti-cancer compounds or, more specifically, 'chemopreventive and therapeutic agents' which might be helpful in the treatment of some forms of cancer. Most of these compounds can be found in a variety of other plants; however, scientists have been able to isolate one unique compound, diomuscipulone, which is found only in the Venus flytrap. Although much further testing is required, it is hoped to have some supplementary therapeutic potential.[17]

The purple pitcher plant (*Sarracenia purpurea*) has a long tradition of use in many indigenous communities. For example, the Cree peoples in northern Québec have traditionally used it to 'treat symptoms

Sarracenia purpurea (purple pitcher plant) at Kew Gardens.

of diabetes'.[18] A recent scientific study has suggested that the plant could show promise as 'an alternative and complementary treatment for diabetic complications'.[19] More generally, the purple pitcher plant has been used as an alternative treatment for chronic pain. One medication, marketed under the name Sarapin, is a suspension made from powdered *Sarracenia purpurea*, which has been used for decades in both human and veterinary medicine. The solution is injected directly into the patient for localized pain relief. Although many have celebrated its positive effects, as yet there has been no scientific evidence of its

Commercial growing of sundew plants.

efficacy. The results of one trial, conducted upon five hundred patients, 'showed no significant differences in the pain relief or duration of significant relief with the addition of Sarapin'.[20]

Several different species of sundew plants (*Drosera*) have been used both in traditional medicine and more recently in homeopathic treatment. The plants have been purported to have strong anti-inflammatory properties, having been listed in various pharmacopeia since at least the 1800s as a treatment for asthma and whooping cough. Today the plant extract is readily available in either a liquid or a tablet form and is used primarily for the treatment of inflammation. Some recent scientific studies have sought to determine if the plants' anti-inflammatory properties might be useful in the treatment of certain types of cancer, but have produced, thus far, uncertain results.[21]

Carnivorous Designs

Carnivorous plants have become iconic design forms that have even made their way into the fashion world. In 2015 Nike launched a newly designed athletic basketball shoe, the 'Kyrie 1 Flytrap'. According to the brand's marketing, 'The Kyrie 1 Flytrap unites [performance] innovation with a colourway story about the Venus flytrap's split-second attack. The shoe's green-and-mango colour scheme is inspired by the hues of a Venus flytrap plant.'[22] In addition to the coloration, the shoes feature a teeth design motif around the front edge of the sole. Another shoe manufacturer, Converse, has released a carnivorous plant-themed sports shoe – the 'Jack Purcell', which features an array of screen-print graphics of both *Nepenthes* and Venus flytraps.

Intriguingly, carnivorous plants have provided inspiration for various industrial design applications. For example, the design and trapping mechanism of the Venus flytrap (which allows for its traps to quickly flip their curvature from a concave to a convex form) is being adopted in the development of experimental devices that could soon

Carnivorous pitcher plant souvenirs – plant pitchers encased in resin – for sale at the Gardens by the Bay, Singapore.

Display of succulent leaved *Pinguicula moranensis* and *P. esseriana*.

have significant benefits in such areas as transport, manufacturing and robotics. In this way, mechanical parts could be made to move with incredible speed and aggregated complexity, without the need for traditional hinge mechanisms.[23] Another design inspiration has come from the slippery surfaces of the peristome (mouth rim) of *Nepenthes* pitchers, which has led to the development of a new surface material so incredibly slippery that virtually nothing can adhere to it.[24] It is anticipated that such surfaces will have wide-ranging applications, from medical instruments to solar panels. By contrast, it has been discovered that the sundew plant's mucilage contains naturally occurring nanoparticles and nanofibres which help to make its glue extra sticky. Researchers hope to be able to synthesize these nano composites to develop new forms of highly effective adhesives.[25]

The 'side-saddle' flowers of *Sarracenia* pitcher plants are frequently used in cut-flower arrangements. Although celebrated for their hanging-petal design, these flowers also have a unique underlying structure so that, even after the flower petals have fallen off, this distinctive structure has strong aesthetic appeal and can remain fresh-looking for several weeks.[26] However, a trend that is becoming increasingly popular in both North America and Japan is for cut-flower arrangements made not from the plants' flowers but from their brightly coloured pitchers. Most commonly, the pitchers of *Sarracenia* and *Darlingtonia* plants are used in these alternative 'cut-flower' arrangements. Frequently the pitchers are mistaken for flowers (a mistake that presumably many insects make too – much to their detriment), so it is not surprising that these have turned up in floral

Artificial plant display featuring synthetic *Darlingtonia* pitchers, Beijing, China.

arrangements. In some instances, the pitchers will be displayed by themselves; in others, they will be combined with traditional cut-flowers. In past decades, the cut-pitchers were often habitat-collected, (picked in the wild) but there are now enough growers cultivating them for sale that a reliable supply is available. Live *Sarracenia* pitcher plants are also being used more frequently in both boutique garden and indoor displays. Artificial carnivorous plants (of either silk or plastic) are also widely available. The most commonly produced are the pitcher plants *Nepenthes*, *Sarracenia* and *Darlingtonia* — synthetic forms that are becoming increasingly popular as decorative items.

six

Collecting, Growing and Conserving Carnivorous Plants

y the late 1800s, carnivorous plants were becoming widely
available and speciality nurseries in England, Europe and
North America had begun to offer an extensive selection. The
most prominent seller of *Nepenthes* in England was the Veitch nursery.
A published review from 1881 proclaimed:

> A visit to Messrs Veitch's collection of *Nepenthes* just now
> will fascinate the plant-lover and stir the pulses of even the
> most indifferent spectator. The free unconstrained way in
> which they fling their branches about, the luxuriance with
> which they hang down their goblets, and their remarkable
> forms and distinct colours leave an impression of as great
> beauty as singularity.[1]

Nepenthes were also available from many North American nurseries
and through their illustrated mail-order catalogues, although these
pitcher plants tended to be much more expensive than the native carnivorous plants also on offer. *John Saul's Catalogue of Plants* (Washington,
DC) for spring 1884 advertised a variety of *Nepenthes*, most of them
at the rather hefty price of $5 each, whereas local species of *Sarracenia*
were advertised at just 30 cents each and Venus flytraps a mere 25
cents.[2]

Pitcher & Manda, a nursery from New Jersey, offered a large selection of *Nepenthes* in its 1892 catalogue, commenting that 'these are

Nepenthes cultivar, showcasing its striking red pitchers.

among the most curious and interesting of cultivated plants, and no collection should be without a few representatives.' These prized plants were priced at $5.[3] The Bellevue Nursery, owned by Wm F. Bassett & Son (also of New Jersey), highlighted the purple pitcher plant, *Sarracenia purpurea*, in its 1897 Catalogue of Plants, noting that 'the strange pitcher like leaves rival the finest Begonias in brilliancy of color and the fine large purple blooms are very beautiful and fragrant.' They could be purchased for 15 cents each, or $1 per dozen.[4]

Another established American nursery, F. Weinberg of New York, which published a periodic catalogue, *Cacti, Novelties, Odd and Rare Plants* also offered a wide range of carnivorous plants. The nursery's 1906 catalogue stated:

Carnivorous plants [are] an interesting family of plants, suitable for the aquatic garden, conservatory, or as house plants, on the window shelf or aquarium. Planted in rather sandy soil, and, if possible, with a little peat and moss they will form into nice plants and shoot up their large, curious flowers annually. They are noted as insect-eating plants, any mosquito or flies, etc., which may alight on them, will hardly ever be able to get away.[5]

The plants on offer included the rather expensive Australian pitcher plant (*Cephalotus follicularis*) at $7.50 each, and various species of *Nepenthes* ranging from $2 to $7.50 per plant. Less expensive were the sundews, which included *Drosera binata* (50 cents to $1), *Drosera capensis* (35 cents to $1) and, slightly more expensive, *Drosera spatulata* (75 cents to $2). *Sarracenia* were advertised as 'handsome natives of our northern bogs and perfectly hardy. Insects which may fly, fall or crawl into them, will not be able to get out of them again.' Available species were offered at modest prices: *Sarracenia drummondii* (25 to 50 cents), *Sarracenia flava*

'Veitch's Nepenthes House', from an 1872 edition of *The Gardeners' Chronicle and Agricultural Gazette*.

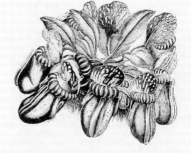

Darlingtonia californea .50 to 1.00   Cephalotus follicularis 5.00 to 7.50

Advertisement for carnivorous plants, from F. Weinberg's *Catalogue of Cacti, Novelties, Odd and Rare Plants* (1906).

(20 to 50 cents) and *Sarracenia purpurea* (15 to 35 cents). Also for sale were established *Darlingtonia californica* (50 cents to $1) and Venus flytraps (*Dionaea muscipula*), '20 cents each or 3 for 25 cents', advertised as 'One of the wonders of vegetable kingdom. The leaves are terminated by clawlike wings, armed with soft, irritable spines. A fly or other insect which may alight on them, will be closed up on them at once.'[6]

By 1908 the huge enterprise of Siebrecht & Son Nurseries had two large nurseries in New York State as well as 'extensive tropical nurseries on our plantations in the West Indies'. These overseas facilities enabled them 'to grow rapidly and to perfection, under most favorable circumstances, many things which are slow and difficult of culture in North America'. The nursery offered 35 varieties of *Nepenthes* (both species and hybrids), ranging in price from $2 to $5, and *Cephalotus follicularis* (the Albany pitcher plant) at $5 each.[7]

E. C. Robbins, a collector and nurseryman from North Carolina, offered what was undoubtedly the widest selection of wholesale carnivorous plants. His 1924 catalogue offered the 'special trade prices' of 1,000 established *Sarracenia flava* for $80, or 1,000 'strong, selected plants' of Venus flytraps for $100.[8] Undoubtedly, at the time many of these plants would have been field collected: 1,000 plants removed from habitat would certainly have made a significant dent in the local population. Owing to decades of such practices, it is estimated that today there are only around 35,000 Venus flytraps left in the wild.[9]

Carnivores as 'Pets'

By the middle of the twentieth century, carnivorous plants, and in particular the Venus flytrap, had gained steadily in popularity. This surge was in part due to the flytrap's regular appearance in popular culture narratives. But they were also being heavily marketed to young people through advertisements in comic books and other youth-orientated publications which, from the 1950s to 1980s, would offer plants for direct sale through the post.

As this marketing effort increased, the manner in which the plants were depicted seemed also to evolve, gradually making a journey from being exotic plants and curiosities, to becoming domesticated pets to which you could hand-feed bits of hamburger. 'The Hungry Plant actually catches and consumes flies!' declared one advertisement (from Deb-Jo Nurseries of New York). 'It will even eat ground meat from your fingers (don't worry, it can't hurt people).' While another ad from the Captain Company of New York affirmed, 'Trap will bite at (but NOT bite off) more than it can chew – such as a finger or a pencil.' Finally, a more recent ad from the 1980s (from Mirobar Sales Corp. of New York), unquestionably intended to evoke an animalistic impression, claimed that as the owner of your new Venus flytrap, you would be able to: 'See how it lures, traps, eats and digests insects up to 20 times it's size! Learn how you can actually train it with a pencil to perform only for you! Feed it raw hamburger from your hand!'

Today, many nurseries and garden centres continue to tap into the pervasive mythology that surrounds these plants. Suppliers will often adorn their stock of Venus flytraps with prominent labels that read 'Fang' or 'Chomper', while *Sarracenia* pitcher plants are frequently sold under such monikers as 'Gobble Guts' or 'Greedy Guts'. In these cases, the intent is not only to emphasize the plants' carnivorous nature, but to imbue them with distinctive pet-like characteristics.

Perhaps one of the best reflections of this theme can be found in a painting by the American artist Will Elder (1921–2008). A painter, illustrator and comic book artist, Elder created a series of humorous paintings that formed part of his 'Grandparents Series'. Featuring an elderly set of grandparents who appeared to be actively encouraging unruly behaviour in their young grandchildren, these paintings would deftly parody characteristic American family life. One image themed around carnivorous plants represents an extremely faithful parody of the iconic American artist Norman Rockwell, who was well known for his gentle yet humorous observations of the idyllic small-town American life of the early twentieth century. Exact in detail, *A Visit to Grandma's* is even signed by the artist in Norman Rockwell's stylistic signature under the pseudonym of 'Rockwill Elder'.

At first glance the subject matter of the painting appears to be quite innocent: children and grandparents are enjoying some quality time in their greenhouse. But on closer inspection it becomes clear that something is amiss. The young boy is, in fact, feeding a grasshopper to an over-sized Venus flytrap, while the grandmother offers encouragement. Then we can see that several other animals are being held in hand, ready to be fed to the hungry plant. The grandmother holds a large frog, which appears to be next in line. The boy also holds

Venus flytraps marketed as 'Fang'.

Will Elder (signed Rockwill Elder), *A Visit to Grandma's*. A parody of a Norman
Rockwell painting, depicting an exaggerated carnivorous pet plant.

a bowl of goldfish, while the young girl holds her pet parakeet. The
family dog sits at their feet looking very guilty, having just torn a piece
of fabric from the grandmother's dress. As of yet, this act appears to
have gone unnoticed. We can only surmise that, when the grandmother discovers what the naughty dog has done, he will serve as the
concluding morsel for their pet plant's supper. Clearly, the Venus
flytrap has become this family's most favoured pet – perhaps it will
soon become their only pet.

Connoisseur Carnivores

Although the 'killer plant' and 'pet-plant' themes persist in the marketing of carnivorous plants, there is an increasing number of collectors and sellers who take the growing and cultivating of these remarkable plants very seriously. In parallel with this increase, there are more species being made available, and more cultivars, hybrids and intriguing mutant varieties becoming accessible.

Pitcher plants, in particular, are very susceptible to cross hybridization, resulting in the emergence of countless new varieties. *Sarracenia*, unlike many other plants, tend to have a fairly predictable hybridized outcome in that their resulting offspring will tend to express equal portions of the characteristics from each of their parent plants.[10] Interestingly, all *Nepenthes* pitcher plant species have been found to have an identical number of chromosomes, an unusual homogeneity that allows for an almost unlimited variety of potential hybrids among these species.[11]

In recent years, probably due to the increase of tissue culture propagation, there has been a dramatic increase in strangely deformed Venus flytraps. Some of these have subsequently been propagated and even given official cultivar names. As with any plant, some growers find mutations intriguing, while other more traditional growers might disparage them. One writer said of these mutated Venus flytraps, 'such abominations deserve no future other than the compost heap'.[12] Others, with entirely different perspectives, are happy to part with many hundreds of dollars for just such an 'abomination'. Some of the more unusual *Dionaea* will have extremely large or very bright red traps. Others might be very distorted, with fused 'teeth' or flattened and stretched open traps that do not shut. Because some of these mutations are so extreme, it has been suggested that plants with non-functional traps should be sprayed occasionally with a foliar fertilizer to make up for their inability to catch insects.[13] *Pinguicula* is another genus of carnivorous plants that

Cultivated sun pitchers (*Heliamphora*).

Display of various carnivorous plants including *Pinguicula* and *Cephalotus*.

has attracted strong interest in this area as many variously deformed specimens have become available.

Although many *Sarracenia* have brightly coloured pitchers, some specimens will lack all pigmentation, their pitchers being a pure green. This albino condition is a fairly rare occurrence caused by their lacking any of the natural red-hued anthocyanin pigments found in other plants. Such specimens are considered highly desirable by some plant collectors.

Growing Carnivorous Plants

Carnivorous plants are spread across multiple genera and families, some growing in very diverse conditions and needing very different growing requirements. But generalized requirements that span across most of the more common varieties include aerated soils, an abundance of water, access to insects (or similar feedings) and good light.

Nearly all carnivorous plants require a well-aerated soil that remains continually moist without becoming waterlogged. Most will do well in a 50/50 mixture of sphagnum peat moss with perlite or clean sand; some will do better with an increased percentage of peat moss. Most require frequent, daily watering, preferably by rain or distilled water (either from steam distillation or reverse osmosis). Tap water (also bottled drinking water and bore-water) tends to be too heavy in minerals and should normally be avoided. This is because most carnivorous plants are accustomed to living in poor growing conditions and are adapted to receiving nutrients through their leaves rather than their roots. Of course, tap water can vary greatly at each location, and in exceptional cases it may be quite low in mineral content and therefore acceptable for use – many growers will test it with a simple water-testing kit. But most would agree that it is far better for the plants to occasionally be given tap water than to dry out.

Many of the more common carnivorous plants benefit from exposure to full sun, particularly Venus flytraps, *Sarracenia* and sundews, all of which generally live in habitats that are wet, but very sunny. The primary exception would be *Nepenthes*, which normally require a more shaded environment.

Plants kept outdoors will normally get all the nutrition that they need from the insects that they capture. Plants kept indoors will probably need additional feeding. Live insects, which can be bought from pet supply stores, might be feasible for just a few plants. A fertilizer pellet can occasionally be dropped into pitcher plants if they are not getting enough natural food. But generally carnivorous plants do not need fertilizer; or should be fertilized in a limited, controlled manner. Careful growers sometimes use a foliar fertilizing technique (spraying the leaves rather than the soil), but this method can vary dramatically from species to species.

Some carnivorous plants, including *Sarracenia, Darlingtonia, Dionaea* and many *Drosera*, will go into a dormant state during the colder winter months, dying back but flourishing again in the spring. They

Assorted carnivorous plants depicted growing in a greenhouse, from Hans Kraemer's
Der Mensch und die Erde (1906).

need this rest period to maintain their health and vigour. Grown in more tropical climates, a dormancy period can be created artificially by placing them in a cool room (or even in a refrigerated space to simulate a winter period).

Dionaea muscipula (Venus flytraps) always prefer full sun. The plants will suffer without enough sunlight, and any new traps will remain stunted. The plants need to be kept moist, and many growers will set their pots in a tray of water. Although the flowers are quite attractive, they divert growth energy away from the trap and most growers trim them. The plants go dormant in the cold winter weather and will do best if allowed to do so – otherwise they are likely to become sickly and die in two to three years. The most common causes of killing a plant are insufficient water or, potentially, too much tap water. Flytraps, in particular, seem to be sensitive to tap water; over time the build-up of minerals can damage the plant. It takes a lot of energy to grow and maintain a trap and each one is only able to snap shut a few times before it dies. So, tempting as it might be, it is probably best to not trigger the traps very often. Kept out of doors, the traps will normally catch all the insects they need, but a plant can be fed a live insect once a month (one fly per plant – not per trap). They grow best in either pure peat moss, or with up to 50 per cent added sand or perlite.

Sarracenia prefer full sun but need to be kept moist. Most growers will keep their plant pots resting in a tray of rain water. In winter, much of the plants' vegetation will die back; the dead pitcher leaves should be trimmed away. Most *Sarracenia* can normally withstand frost. Even though the plants will remain mostly dormant during winter, the soil should be kept moist. However, if the plants are being grown in very warm and tropical climates or are being grown indoors, it is recommended that the rhyzomes (the underground bulb-like portion of the plant) be dug out and placed in a plastic bag in the refrigerator for a few months of the year. This will simulate a cold winter period and encourage a beneficial dormancy period. Many recommend a one-part sand to five-parts peat moss soil mixture, but some growers will

increase the sand content to a simple mixture of 50 per cent sand or perlite and 50 per cent peat moss. *Sarracenia purpurea* is one of the easiest of the American pitcher plants to grow. It is a very hardy plant that can withstand a wide range of temperatures. *Sarracenia leuxophylla* (the white trumpet pitcher) and its hybrids are some of the most spectacular pitcher plants. In contrast to the rather squat stature of the purple pitcher, these magnificent pitchers can grow up to a metre tall. The flowers are generally bright red in colour. Provided the basic care requirements of *Sarracenia* are followed, then these plants will also be easy to look after.

Nepenthes normally need a semi-shaded location, as they do not do well in bright sun or frost. The highland species can normally withstand cooler climates than the lowland species. The plants should be kept moist, but in a pot that has very good drainage; and, of course, rain water is best. Many growers will also mist spray their plants regularly to emulate the high humidity that they are accustomed to in their natural habitat. As the pitchers die back they can be trimmed off. There are a number of hybrid species that are quite hardy and, depending upon your local climate, these will do quite nicely outdoors if protected under a tree or an overhang such as a patio cover.

Sundews, as the name suggests, prefer full sun. These plants also need to be kept moist and most growers will place the plant pots in a tray of water. Some, but not all, sundews will go dormant in winter. Cape sundews (*Drosera capensis*), native to South Africa's Cape region, are known as the easiest of the sundews to grow (and in fact, the easiest of all carnivorous plants). In warm climates, this species will flourish year-round and does not require a winter dormancy period. If grown in cooler climates it can usually withstand some freezing; but in cold climates it will normally die back to its roots and then flourish again in spring. The plant will also do nicely year-round on a sunny windowsill. The strikingly red-coloured, fork-leafed *Drosera binata*, which is native to Australia, is also relatively easy to grow. Although not as

Collection of *Sarracenia* cultivars.

Cape sundews (*Drosera capensis*), native to South Africa's Cape region, are regarded as one of the easiest species of carnivorous plants to grow.

robust as *D. capensis* it can withstand colder weather in winter and will die back to its roots. Similarly, *D. rotundifolia*, native to the UK, is also a fairly hardy plant. It enjoys full sun in summer and is used to a winter dormancy.

Carnivorous Plant Conservation

Many carnivorous plants are threatened or considered vulnerable in habitat. This is owing to both the destruction of their natural environment and from the over-collecting of plants in the wild. In Borneo, populations of highly prized *Nepenthes rajah* and *Nepenthes northiana* were decimated by collectors several decades ago, when a habitat-collected *N. rajah* would fetch over $1,000.[14] There was also very strong demand for field-collected *Sarracenia*, Venus flytraps and many other carnivorous plants.

Fortunately, the unscrupulous wholesale collecting of wild plants has diminished in recent years. One reason for this is the creation of numerous local laws which now make it illegal to collect any plant materials from these protected species without a permit. Currently in the U.S., one can be fined up to $1,000 for each Venus flytrap that is removed from habitat. In a local effort to enforce this, the North Carolina Department of Agriculture has been routinely painting UV dye onto the wild plants. The dye is invisible until a UV light is shone on it. This way, inspectors are able to determine whether a plant has been field-collected or has been cultivated.[15]

The decline in field-collecting can also be attributed to the creation of the Convention on International Trade in Endangered Species of Wild Fauna and Flora (CITES), agreed in 1973 and in force from 1975. This major international treaty aims to protect endangered fauna and flora, including threatened and endangered carnivorous plants. The treaty has now been signed by more than 130 countries; it maintains a list of all internationally protected specimens divided into two main categories. Species that are listed in Appendix I are heavily protected and described as being at serious risk

Darlingtonia californica in habitat, northern California.

Display of a variety of carnivorous *Nepenthes* pitcher plants.

from commercial trade. Plants listed in Appendix II are considered to be slightly less at risk but are nevertheless very strictly regulated. The Venus flytrap is listed in Appendix II, as are all species of *Nepenthes* – the exceptions are *N. rajah* and *N. khasiana*, which are in Appendix I, meaning that international trade of these field-collected specimens is totally banned. While most *Sarracenia* appear in Appendix II, *S. oreophila*, *S. rubra* ssp. *albamensis* and *S. rubra* ssp. *jonesii* are also in Appendix I.

Today many of these species are available from tissue culture, which is a technique used to clone plants from very small tissue samples. This allows for thousands of plants to be created very quickly, and from a very small portion of a plant – making wild-collecting of most species an almost obsolete practice. However, there are still localized problems where collectors will seek well-established specimens, or more recently discovered species that have not yet been widely tissue-cultured.

Although CITES, local regulations and the increasing availability of tissue-cultured specimens have diminished demand for field-collected plants, habitat destruction has emerged as a far greater problem in recent years. As Richard Mabey has pointed out regarding laws that protect the Venus flytrap, 'In the past, flytrap populations may have suffered from over collecting. Currently U.S. law protects the plants in the wild, with collection punishable by fines of up to a thousand dollars per plant. Ironically, habitat can be drained and thousands of plants killed without any fines at all.'[16]

A similar situation exists in Borneo, where individual *Nepenthes* are highly protected, but huge areas of *Nepenthes* forest habitats (including millions of individual pitcher plants) are regularly demolished for logging and agriculture. In Borneo, millions of hectares of pristine *Nepenthes* habitat have been converted to palm-oil plantations in recent years.[17]

Many carnivorous plants, particularly pitcher plants (*Nepenthes* and *Sarracenia*), support a wide range of animals, including insects, frogs and bats. The loss of these plants could therefore spell devastation for many animal species. For example, if *Nepenthes hemsleyana* were to become extinct, then the Hardwicke's woolly bat (*Kerivoula hardwickii*), which relies almost exclusively on these pitchers for their sleeping quarters, would also greatly suffer. And, because this particular pitcher plant relies almost entirely on the bat's guano for its nutrition, it would suffer in turn if the bat population were to plummet.

Although saving individual plants is important, it is impossible to replicate natural animal/plant mutualism in a greenhouse. It is also very difficult to preserve such relationships in small pocket-reserves – expansive natural habitats are typically needed for this. Fortunately, there have been some recent governmental efforts to protect natural habitats, and as a result some sizeable areas have been set aside as reserves. For example, in 2007 the various nations of Borneo signed a far-reaching agreement to protect vast areas of wilderness. The agreement also seeks to establish more state and national parks and to ensure that these are linked by protected forested corridors.[18]

Many North American carnivorous plants rely upon regular seasonal fires for their survival. Carnivorous plants are not very strong competitors for the limited resources of nutrients and sunlight; they rely upon fires to reduce the competition from invasive species and other more aggressive plants.[19] The human tendency, particularly near built-up areas, is to suppress such fires. Banning fires in these areas causes harm to the carnivorous plant populations. By contrast, fire has become a serious problem in the jungle forests of Borneo. Owing to increased logging and forest destruction, drought and global warming, fires are occurring with increasing frequency. In recent years, fires have destroyed huge populations of *Nepenthes* as well as many other flora and fauna.[20]

In some areas, populations of waterwheel plants (*Aldrovanda vesiculosa*) are being severely threatened by habitat destruction. When a particular plant population is endangered in this way it is a common practice for scientists and conservationists to collect seeds from the species, placing them into a seed bank to ensure against total extinction. In the case of the waterwheel plant, the seeds are proving to be very difficult to store. Kept at normal temperatures, they are highly

Ernst Haeckel, illustration of Nepenthaceae from *Kunstformen der Natur* (1904).

prone to fungal infestations; but placed into deep freeze, even for short periods, the seeds have all failed to germinate.[21]

The two main approaches to carnivorous-plant conservation are 'in situ', the protection of plants in their native habitat (arguably the most effective), and 'ex situ', the cultivation of plants away from their habitat, in botanic gardens or private collections. A number of conservationists have cautioned that relying purely on the latter method will have a negative long-term effect on the genetic variability of the plants. Unfortunately, there are many instances in which ex-situ conservation is the only option, and some significant and very vital conservation work has been achieved by botanical organizations as well as by private growers. To this end, carnivorous plant expert

Nepenthes in habitat, Madagascar.

Stewart McPherson has set up the Ark of Life Foundation, which focuses extensively on the protection and ex-situ preservation of *Nepenthes* species. Similarly, the North American *Sarracenia* Conservancy (NASC) is a long-established group that seeks to protect, conserve, propagate and restore the natural habitat of *Sarracenia*. The organization also has many members who are part of its Grower Committee, dedicated to the ex-situ cultivation of particular species of *Sarracenia*:

> The goals of the North American *Sarracenia* Conservancy (NASC) Grower Committee are to propagate collected plant materials, including seeds for the purposes of reintroduction and maintaining a permanent record of the genetic diversity of the genus *Sarracenia* and companion plants . . . This is not a single collection of plants in one central location. Rather, the NASC collection is held by many NASC growers throughout the U.S. By maintaining individual types of collection plants in many places, it is hoped that the risk of catastrophic loss for whatever reason will be minimized.[22]

Other conservation work carried out by NASC includes relocating populations of threatened plants and re-introducing plants into areas previously populated by *Sarracenia*. The group performs salvage operations, saving plants from areas that are due to be destroyed for roadworks and urban development. Importantly, they have initiated essential prescribed burning operations in order to facilitate the growth of native populations.

Carnivorous Plant Societies

In recent decades, a number of carnivorous plant societies have formed all over the world. The world's oldest carnivorous plant society was founded in Japan in 1949. In 1972 the International Carnivorous Plant

Overleaf: Assorted *Sarracenia* pitcher plants.

Sarracenia's flower structure, after its petals have dropped off.

Society (which is based in California) was founded. The Carnivorous Plant Society of the UK was formed in 1978; the French society, Association Francophone des Amateurs de Plantes Carnivores, commenced in 1983; the German society, Gesellschaft für Fleisch-fressende Pflanzen, in 1984; the Victorian Carnivorous Plant Society in Melbourne, Australia, also in 1984; and the Italian society, Associazione Italiana Piante Carnivore, in 1998.

People join established carnivorous plant societies for a wide range of reasons, but many do so in order to gain access to plants that cannot be purchased at the local nursery, to learn more about carnivorous plants, and to meet people who share a similar interest. Members range from highly specialized botanists to hobbyists who simply like carnivorous plants. Societies hold regular meetings that feature guest speakers, discussions, plant sales and exhibits. Larger societies also hold conferences that spotlight nationally and internationally recognized speakers. Such conferences might also incorporate exhibitions, workshops, plant sales and excursions to significant gardens and collections. Each year, societies will hold carnivorous plant shows and competitions, which help foster excellence in the growing and presentation of spectacular plant specimens. These competitions also serve to educate the public about carnivorous plants and to attract new members.

Another important role of these societies, particularly in recent decades, has been to facilitate the conservation, cultivation and scientific study of carnivorous plants. Many societies will sponsor field studies, plant rescue operations and other conservation efforts.

In recent years, the internet has greatly expanded the popularity of carnivorous plants. Many carnivorous plant societies now have a strong online presence, and many other forums and networks have sprung up to allow people to showcase their plant collections, buy and sell plants, provide growing tips and post images requesting help in the identification or care of particular specimens.

Carnivorous Plant Futures

In a 1909 issue of *Scientific American*, there appeared a lengthy article in which the author speculated on the future of carnivorous plants. S. Leonard Bastin describes how, if carnivorous plants were to follow their current evolutionary trajectory, they are likely to grow to enormous size and become much more 'deadly':

> It is a fact recognized by botanists as beyond dispute that the carnivorous habit among plants is more widespread than it was formerly supposed to be. The specialized sundews (*Drosera*) are but the advance guard of a large army of species which depend for their existence more or less upon the absorption of animal salts through their foliage. There is

no gainsaying the statement recently put forward by more than one scientist, that the tendency to rely upon a carnivorous diet is on the increase. Of course, this is only in a line with the simplest evolutionary principle. It is possible to trace the steps by which the highest types of species, which seize and hold their prey, such as the Venus fly trap (*Dionea*), have been evolved from those which merely capture their victims by the use of an adhesive fluid, such as the fly catcher of Portugal (*Drosophyllum*) . . . It is a startling conception that in ages to come the plant world as a whole may become so advanced in carnivorous tastes as to be a real menace to animal creation.[23]

The author then details each of the most common groupings of carnivorous plants: sundews, Venus flytraps, butterworts, bladderworts and pitcher plants (including the Australian, *Cephalotus*), describing firstly how these plants in their present form go about capturing insects; then amplifying these attributes, describing how they might capture large animals, even humans, if they were to evolve to a mammoth size. For example, should one fall into 'the pitcher of the *Cephalotus*, escape would be possible only with a friendly assistant at hand'. Furthermore, that giant *Pinguicula* might develop a taste for goats and cattle, or bladderworts might begin to eat crocodiles. Most troubling, he notes:

> Far more dreadful than any of the plants described above would be the Venus flytrap of the future. This plant would be a vegetable terror . . . Any unfortunate man who should chance to stumble into one of these leaves would be speedily crushed to death by the steady pressure of the inclosing sides.[24]

The author then concludes this 'scientific' article with a touch of cautious optimism: 'It may be that the condition of man himself will have changed considerably by the time he is called upon to face these aggressive plants. It is to be hoped that this may be so, otherwise the outlook for the human race is distinctly disquieting.'[25]

More realistically, we could invert this turn-of-the-century speculation and argue that if we do not look after the habitats of these remarkable carnivores, the future of these plants – not that of humanity – 'is

Hanging pitchers
of a *Nepenthes* cultivar.

197

A highly speculative illustration, published in 1909 in *Scientific American*, depicting what an Albany pitcher plant (*Cephalotus follicularis*) might look like in the future.

distinctly disquieting'. Hopefully, as more of us recognize the beauty and extreme complexity of these carnivorous wonders of nature, how intricately they are intertwined with the existence of a great number of other creatures, the more we will begin to care for them. In doing so, we are likely to continue to discover many new species of carnivores, and also learn a lot more about the ones we already enjoy.

Timeline

❦

1200s	Sundews (*Drosera*) and butterworts (*Pinguicula*) are used medicinally for their antibacterial properties
1554	Publication of *Cruyde Boeck*, an early 'herbal book' by Rembertus Dodonaeus, which features the earliest illustration of a sundew plant
1576	Mathias de L'Obel publishes the book *Nova stirpium adversaria*, which features what are likely the first illustrations of *Sarracenia* pitcher plants
1578	Henry Lyte writes about the sundew plant in his book *New Herball*
1658	The French governor of Madagascar, Etienne de Flacourt, describes the local pitcher plants in his text *Histoire de la Grande Isle de Madagascar*, which also includes the first published illustration of this plant. He refers to the species as *Anramatico* (now known specifically as *Nepenthes madagascariensis*)
1682	Renegade naturalist Nehemiah Grew, in his book *The Anatomy of Plants*, suggests that plants, on a fundamental level, are quite similar to animals
1686	John Ray first writes about *Nepenthes* in his book *Historia plantarum*
1759	The Venus flytrap is first discovered by Arthur Dobbs, Governor of the Colony of North Carolina

1768	John Ellis publishes his first article on the Venus flytrap
1787	*Curtis's Botanical Magazine* commences publication under founding editor William Curtis. It contains many articles and illustrations featuring carnivorous plants – particularly *Nepenthes*
1789	Joseph Banks introduces *Nepenthes* to England – a specimen of *N. distillatoria* to the Royal Botanic Gardens at Kew
1830s	Dr Nathaniel Bagshaw Ward invents the Wardian case
1860	Charles Darwin commences his study of carnivorous plants
1865	At the recommendation of Queen Victoria, Newfoundland adopts the carnivorous pitcher plant *Sarracenia purpurea* as its provincial flower. Featuring this plant on its 1-cent coin, Newfoundland produces the world's first currency to depict a carnivorous plant
1874	Joseph Hooker publishes the ground-breaking article 'The Carnivorous Habits of Plants', which focuses primarily on the pitcher plants: *Nepenthes*, *Sarracenia* and *Darlingtonia*
1875	Charles Darwin's *Insectivorous Plants* is published
1880	Arthur Conan Doyle publishes his seminal short story 'The American's Tale' about man-eating Venus flytraps
1882	John Burdon Sanderson publishes the book *Excitability of Plants*, which details how electrical pulses are used to trigger the closure of the leaves of the Venus flytrap
1885	Mary Treat, a well-respected naturalist, publishes her best-selling book *Home Studies in Nature*, which details many of her observations of *Sarracenia* and *Utricularia*
1949	The Carnivorous Plant Society of Japan is founded. This was the world's first established carnivorous plant society

1951	John Wyndham's *The Day of the Triffids* is published
1960	The movie *The Little Shop of Horrors* (directed by Roger Corman) is released
1962	Movie version of *The Day of the Triffids* (directed by Steve Sekely) is released
1970	*Body of the Prey* (aka *Venus Flytrap*), directed by Norman Earl Thomson and written by Ed Wood, is released in cinemas
1972	The ICPS (International Carnivorous Plant Society) is founded in California
1973	The CITES (Convention on International Trade in Endangered Species of Wild Fauna and Flora) international treaty is created. Many species of carnivorous plants are included in the listings
1978	The Carnivorous Plant Society of the UK is formed
1981	A BBC six-part miniseries production of *The Day of the Triffids*, directed by Ken Hannam, is first broadcast
1982	The off-Broadway stage musical *Little Shop of Horrors* opens in New York city
1983	The French carnivorous plant society, Association Francophone des Amateurs de Plantes Carnivores, is founded
1984	The Victorian Carnivorous Plant Society is formed in Melbourne, Australia
1984	The German carnivorous plant society, Gesellschaft für Fleischfressende Pflanzen, is founded
1986	The movie *Little Shop of Horrors*, directed by Frank Oz, is released in cinemas. This is an adaptation of the stage musical and of the original 1960 movie
1991	The animated television series *Little Shop* is first broadcast

1998	The Italian carnivorous plant society, Associazione Italiana Piante Carnivore, is founded
2001	Simon Clark's *The Night of the Triffids*, a sequel to the original book *The Day of the Triffids*, is published
2005	The North American *Sarracenia* Conservancy (NASC) is founded
2009	A new two-part telemovie BBC production of *The Day of the Triffids*, directed by Nick Copus, is first broadcast
2011	Stewart McPherson launches the Ark of Life Foundation for the ex-situ preservation of carnivorous plants
2012	*Philcoxia minensis* is proven to be a carnivorous plant. Soon after, additional species from the genus are also determined to be carnivorous
2013	Malaysian hit children's television series *Upin and Ipin* devote an episode to learning about *pulut periuk kera* (pitcher plant rice)
2015	Nike launches the Kyrie 1 Flytrap athletic basketball shoe, modelled after the Venus flytrap

References

1 The Natural History of Carnivorous Plants

1 David E. Jennings and Jason R. Rohr, 'A Review of the Conservation Threats to Carnivorous Plants', *Biological Conservation*, CXLIV (2011), p. 1357.
2 David H. Benzing, *Air Plants: Epiphytes and Aerial Gardens* (Ithaca, NY, and London, 2012), p. 185.
3 James Pietropaolo and Patricia Pietropaolo, *Carnivorous Plants of the World* (Portland, OR, 1986), p. 6.
4 Ibid.
5 Ibid., p. 7.
6 Jim D. Karagatzides, Jessica L. Butler and Aaron M. Ellison, 'Pitcher Plant *Sarracenia purpurea* and the Inorganic Nitrogen Cycle', in *Plant Physiology*, ed. Philip Stewart and Sabine Globig (Toronto and New York, 2012), p. 17.
7 Ibid., p. 16.
8 Thomas C. Gibson and Donald M. Waller, 'Evolving Darwin's "Most Wonderful" Plant: Ecological Steps to a Snap-Trap', *New Phytologist*, CLXXXIII (2009), pp. 575–87.
9 Nigel Hewitt-Cooper, *Carnivorous Plants: Gardening with Extraordinary Botanicals* (Portland, OR, 2016), p. 32.
10 Ibid., p. 38.
11 Stephen E. Williams and Siegfried R. H. Hartmeyer, 'Prey Capture by *Dionaea muscipula*. A Review of Scientific Literature with Supplementary Original Research', *Carnivorous Plant Newsletter: Journal of the International Carnivorous Plant Society*, XLVI/2 (June 2017), pp. 44–61.
12 Alexander G. Volkov et al., 'Electrical Memory in Venus Flytrap', *Bioelectrochemistry*, LXXV (2009), pp. 142–7.
13 Malcolm Wilkins, *Plantwatching: How Plants Remember, Tell Time, Form Relationships and More* (New York, 1988), p. 139.
14 Volkov et al., 'Electrical Memory in Venus Flytrap', pp. 142–7.
15 Ibid.

16 Vladislav S. Markin and Alexander G. Volkov, 'Morphing Structures in the Venus Flytrap', in *Plant Electrophysiology Signaling and Responses*, ed. Alexander G. Volkov (New York, 2012), p. 1.

17 Ibid., p. 23.

18 Tim Bailey and Stewart McPherson, *Dionaea: The Venus's Flytrap* (Poole, 2012), p. 105.

19 Hewitt-Cooper, *Carnivorous Plants*, p. 97.

20 Edward E. Farmer, *Leaf Defence* (Oxford, 2014), p. 93.

21 Wilhelm Barthlott et al., *The Curious World of Carnivorous Plants: A Comprehensive Guide to Their Biology and Cultivation* (Portland, OR, and London, 2007), p. 117.

22 Ibid., p. 136.

23 Ibid., p. 137.

24 Ibid., p. 105.

25 Lubomír Adamec, 'Ecophysiological Investigation on *Drosophyllum Lusitanicum*: Why Doesn't the Plant Dry Out?', *Carnivorous Plant Newsletter: Journal of the International Carnivorous Plant Society*, XXXVIII/3 (2009), pp. 71–4.

26 Peter D'Amato, *The Savage Garden: Cultivating Carnivorous Plants*, revd edn (Berkeley, CA, 2013), p. 219.

27 Barthlott et al., *The Curious World of Carnivorous Plants*, p. 120.

28 Tan Wee Kiat and Amy Sabrielo, *Jack and the Carnivorous Pitcher Plant* (Singapore, 1999), p. 12.

29 Hewitt-Cooper, *Carnivorous Plants*, p. 200.

30 Naoya Hatano and Tatsuro Hamada, 'Proteome Analysis of Pitcher Fluid of the Carnivorous Plant *Nepenthes alata*', *Journal of Proteome Research*, VII/2 (2008), p. 815.

31 Barthlott et al., *The Curious World of Carnivorous Plants*, p. 157.

32 Ibid., p. 158.

33 Anthea Phillipps, Anthony Lamb and Ch'ien C. Lee, *Pitcher Plants of Borneo*, 2nd edn (London and Borneo, 2008), p. 57.

34 Hewitt-Cooper, *Carnivorous Plants*, p. 191.

35 Carnivorous plant expert and author Peter D'Amato is credited with coining the common name, corkscrew plant.

2 More than Just a Meal

1 Stewart McPherson, *Glistening Carnivores: The Sticky-leaved Insect-eating Plants* (Poole, 2008), p. 59.

2 Stewart McPherson, *Carnivorous Plants and their Habitats*, vol. II (Poole, 2010), p. 982.

3 McPherson, *Glistening Carnivores*, p. 61.

4 Wilhelm Barthlott et al., *The Curious World of Carnivorous Plants: A Comprehensive Guide to Their Biology and Cultivation* (Portland, OR, and London, 2007), p. 176.

5 Anthea Phillipps, Anthony Lamb and Ch'ien C. Lee, *Pitcher Plants of Borneo*, 2nd edn (London and Borneo, 2008), p. 62.

6 Ibid., p. 58.
7 Ibid., p. 64.
8 Ibid.
9 Ibid.
10 Ibid., p. 66.
11 Ibid., p. 58.
12 Ibid., p. 59.
13 Melinda Greenwood et al., 'A Unique Resource Mutualism between the Giant Bornean Pitcher Plant, Nepenthes rajah, and Members of a Small Mammal Community', *PLoS ONE*, VI/6 (2011), pp. 3–5.
14 Phillipps et al., *Pitcher Plants of Borneo*, p. 40.
15 Michael G. Schoner et al., 'Bats are Acoustically Attracted to Mutualistic Carnivorous Plants', *Current Biology*, XXV (July 2015), pp. 1911–16.
16 Ibid.
17 Ibid.
18 Charles Darwin, *Insectivorous Plants* (London, 1875), p. 110.
19 Ibid., p. 316.
20 Andrew Wilson, 'The Inner Life of Plants', *Gentleman's Magazine*, CCLV (1883), p. 232.
21 Phillipps et al., *Pitcher Plants of Borneo*, p. 79.
22 Tim Bailey and Stewart McPherson, *Dionaea: The Venus's Flytrap* (Poole, 2012), p. 80.
23 Pedro Nájera Quezada, 'Carnivorous Xeric Flora in San Luis Potosí, México', *Xerophilia*, III/3 (October 2014), pp. 4–16.
24 Barthlott et al., *The Curious World of Carnivorous Plants*, p. 101.
25 John Dawson and Rob Lucas, *The Nature of Plants: Habitats, Challenges, and Adaptations* (Melbourne, 2005), p. 262.
26 John Brittnacher, 'Murderous Plants', *Carnivorous Plant Newsletter: Journal of the International Carnivorous Plant Society*, XL/1 (March 2011), p. 17.
27 McPherson, *Glistening Carnivores*, pp. 47–9.
28 Douglas W. Darnowski, *Triggerplants* (Sydney, 2002), p. 71.

3 A Remarkable Discovery

1 Wilhelm Barthlott et al., *The Curious World of Carnivorous Plants: A Comprehensive Guide to Their Biology and Cultivation* (Portland, OR, and London, 2007), p. 149.
2 Anthea Phillipps, Anthony Lamb and Ch'ien C. Lee, *Pitcher Plants of Borneo*, 2nd edn (London and Borneo, 2008), p. 68.
3 Ibid.
4 Translated from the Latin and quoted ibid., p. 69.
5 Henry Lyte, *A New Herball* (1578), quoted in Nigel Hewitt-Cooper, *Carnivorous Plants: Gardening with Extraordinary Botanicals* (Portland, OR, and London, 2016), p. 14.
6 Roy Vickery, *A Dictionary of Plant-lore* (Oxford, 1995), p. 362.
7 Anne Pratt, *The Flowering Plants of Great Britain* (London, 1855), p. 188.

8 Nehemiah Grew, *The Anatomy of Plants* (London, 1682).

9 Emma Darwin, quoted in Stewart McPherson, *Carnivorous Plants and their Habitats*, vol. 1 (Poole, 2010), p. 36.

10 T. H. Huxley, 'On the Border Territory Between the Animal and Vegetable Kingdom', *Macmillan's Magazine: The Popular Science Monthly*, VIII/8 (1 April 1876), p. 641.

11 Arthur Dobbs, quoted in Tim Bailey and Stewart McPherson, *Dionaea: The Venus's Flytrap* (Poole, 2012), p. 17.

12 Ibid.

13 Richard Mabey, *The Cabaret of Plants: Botany and the Imagination* (London, 2015), p. 188.

14 Daniel L. McKinley, '"Wagish Plant as Wagishly Described", John Bartram's Tipitiwitchet: A Flytrap, Some Clams and Venus Obscured', in E. Charles Nelson, *Aphrodite's Mousetrap: A Biography of Venus's Flytrap* (Aberystwyth, 1990), pp. 130–32.

15 Peter Collinson, quoted in Bailey and McPherson, *Dionaea: The Venus's Flytrap*, p. 21.

16 John Ellis, '*Dionaea muscipula*', *The British Evening Post* (1 September 1768), quoted in Nelson, *Aphrodite's Mousetrap*, p. 38.

17 Quoted in Nelson, *Aphrodite's Mousetrap*, p. 46.

18 John Ellis, 'A Botanical Description of a New Sensitive Plant, Called *Dionaea musciopula*: or, Venus's Fly-trap', in *Directions for Bringing over Seeds and Plants, from the East Indies and other Distant Countries, in a State of Vegetation: Together with a Catalogue of Such Foreign Plants as are Worthy of Being Encouraged in Our American Colonies, for the Purposes of Medicine, Agriculture, and Commerce* (London, 1770), p. 37.

19 Carolus Linnaeus, quoted in McPherson, *Carnivorous Plants and their Habitats*, vol. 1, p. 15.

20 Erasmus Darwin, quoted in McPherson, *Carnivorous Plants and their Habitats*, vol. 1, p. 16.

21 The American Tract Society, *Travellers' Wonders*, III (1830), p. 15.

22 Etienne de Flacourt, quoted in Phillipps et al., *Pitcher Plants of Borneo*, p. 3.

23 Phillipps et al., *Pitcher Plants of Borneo*, p. 67.

24 Carolus Linnaeus, translated and quoted in Harry James Veitch, 'Nepenthes', *Journal of the Royal Horticultural Society*, XXI (7 September 1897), p. 229.

25 Phillipps et al., *Pitcher Plants of Borneo*, p. 7.

26 Sydney H. Vines, 'The Physiology of Pitcher-plants', *Journal of the Royal Horticultural Society*, XXI (1897), p. 96.

27 J. E. Smith, *Introduction to Physiological and Systematic Botany*, 2nd edn (1809), p. 195. Quoted in Vines, 'The Physiology of Pitcher-plants', p. 98.

28 Charles Darwin, *Insectivorous Plants* (London, 1875), p. 1.

29 Quoted in Stewart McPherson, *Carnivorous Plants and their Habitats* (Poole, 2010), vol. 1, p. 35.

30 Darwin, *Insectivorous Plants*, p. 294.

31 McPherson, *Carnivorous Plants and their Habitats*, vol. 1, p. 42.

32 Pratt, *Flowering Plants of Great Britain*, p. 188.

33 J. G. Hunt, 'Natural History Studies', *Friends' Intelligencer*, XXXIX/1 (1882), p. 10, quoted in Tina Gianquitto, 'Criminal Botany Progress, Degeneration, and Darwin's Insectivorous Plants', in *America's Darwin: Darwinian Theory and U.S. Literary Culture*, ed. Tina Gianquitto and Lydia Fisher (Athens, GA, 2014).

34 Reginald Farrer, *Alpines and Bog-plants* (London, 1908), pp. 245–6.

35 Ibid.

36 Colin Clout, *Colin Clout's Calendar: The Record of a Summer, April–October* (London, 1883), p. 138.

37 Cesare Lombroso, *Criminal Man* [1884], ed. and trans. M. Gibson and N. H. Rafter (Durham, NC, 2006), p. 167. Also quoted in Dawn Keetley, 'Introduction: Six Theses on Plant Horror; or, Why Are Plants Horrifying?', in *Plant Horror: Approaches to the Monstrous Vegetal in Fiction and Film*, ed. Dawn Keetley and Angela Tenga (London, 2016), p. 17.

38 'Mars Peopled by One Vast Thinking Vegetable', *Salt Lake Tribune* (13 October 1912).

39 'Carnivorous Plants Killed by Indigestion', *Omaha Daily Bee* (11 December 1904), p. 34.

40 Alice Lounsberry, 'Plants that Set Traps', *New York Tribune* (20 January 1907) p. 15.

41 Rene Bache, 'Cleverest of All Plants', *Ogden Standard* (23 March 1918), p. 37.

42 'The Animalism of Plants', *Scientific American*, XXXIII/1 (1875).

4 Attack of the Killer Plants

1 The Editor, 'Editorial', in *Startling Stories* (September 1951), p. 123.

2 Cesare Lombroso, *Criminal Man* [1884], ed. and trans. M. Gibson and N. H. Rafter (Durham, NC, 2006).

3 Dawn Keetley, 'Introduction: Six Theses on Plant Horror; or, Why Are Plants Horrifying?', in *Plant Horror: Approaches to the Monstrous Vegetal in Fiction and Film*, ed. Dawn Keetley and Angela Tenga (London, 2016), p. 6.

4 'Some Remarkable Trees', *Topeka Daily State Journal: Saturday Evening* (4 October 1913), p. 18.

5 'Sacrificed to a Man-eating Plant', *Ogden Standard-examiner* (Sunday Morning, 26 September 1920).

6 Arthur Conan Doyle, 'The American's Tale', *London Society*, XXXVIII (December 1880), p. 44.

7 Ibid.

8 Howard R. Garis, 'Professor Jonkin's Cannibal Plant', *Argosy* (August 1905).

9 René Morot, 'Drosera Cannibalis', *The Living Age* (25 February 1922).

10 Ibid.

11 H. Thompson Rich, 'The Beast Plants', *Famous Fantastic Mysteries*, II/1 (April 1940) pp. 66–77.

12 Barry Langford, 'Introduction', in John Wyndham, *The Day of the Triffids* (London, 1999), p. ix.

13 Constantine Nasr, ed., *Roger Corman Interviews* (Jackson, MS, 2011), p. 102.

14 'The Monster of Frankenstein and the Plant', *The Monster of Frankenstein*, XXXIII (New York, 1954), pp. 23–31.

15 Marc Jensen, '"Feed Me!": Power Struggles and the Portrayal of Race in Little Shop of Horrors', *Cinema Journal*, XLVIII/1 (Autumn 2008), p. 57.

16 *Pokemon – Gotta Catch 'em All: Deluxe Essential Handbook* (Sydney, 2015), p. 57.

17 Ibid., p. 405.

5 Magnificent Carnivores

1 Gina la Morte, 'Earth's Gold', *Boho*, X (Spring 2011), p. 28.

2 See www.madelinevonfoerster.com, accessed 1 February 2017.

3 Madeline von Foerster, correspondence with author, September 2017.

4 Ibid.

5 Ibid.

6 Ibid.

7 See www.noirscapes.com, accessed 1 February 2017.

8 See www.evankolker.com, accessed 10 February 2017.

9 See www.jasongamrathglass.com, accessed 10 February 2017.

10 See http://dancorson.com, accessed 10 February 2017.

11 Ibid.

12 See www.paulhillsculpture.com, accessed 10 February 2017.

13 Anthea Phillipps, Anthony Lamb and Ch'ien C. Lee, *Pitcher Plants of Borneo*, 2nd edn (London and Borneo, 2008), p. 69.

14 Wilhelm Barthlott et al., *The Curious World of Carnivorous Plants: A Comprehensive Guide to Their Biology and Cultivation* (Portland, OR, and London, 2007), p. 149.

15 Phillipps et al., *Pitcher Plants of Borneo*, p. 70.

16 See www.carnivora.com, accessed 1 August 2017.

17 François Gaascht, Mario Dicato and Marc Diederich, 'Venus flytrap (*Dionaea muscipula* Solander ex Ellis) contains Powerful Compounds that Prevent and Cure Cancer', *Frontiers in Oncology*, III (August 2013), pp. 1–18.

18 Cory S. Harris et al., 'Characterizing the Cytoprotective Activity of *Sarracenia purpurea* L., a Medicinal Plant that Inhibits Glucotoxicity in PC12 cells', *BMC Complementary and Alternative Medicine*, XII/245 (2012), pp. 1–10.

19 Ibid.

20 Kavita N. Manchikanti et al., 'A Double-blind, Controlled Evaluation of the Value of Sarapin in Neural Blockade', *Pain Physician*, VII (2004), pp. 59–62.

21 N. B. Ghate et al., 'Sundew Plant, a Potential Source of Anti-inflammatory Agents, Selectively Induces G2/M Arrest and Apoptosis in MCF-7 Cells through Upregulation of p53 and Bax/Bcl-2 Ratio', *Cell Death Discovery*, II (2016), pp. 1–10.

22 'Kyrie 1 Flytrap Basketball Shoe Captures Deceptive Quickness', www.nike.com, 15 February 2015.

23 Nakul Prabhakar Bendea et al., 'Geometrically Controlled Snapping Transitions in Shells with Curved Creases', *Proceedings of the National Academy of Sciences of the United States of America*, CXII/36 (2015), pp. 11175–80; Mohsen Shahinpoor, 'Biomimetic Robotic Venus Flytrap (*Dionaea muscipula Ellis*) made with ionic polymer Metal Composites', *Bioinspiration and Biomimetics*, VI (2011), pp. 1–11.

24 Ed Yong, 'Killer Plant Super-slippery Material that Repels Everything', www.phenomena.nationalgeographic.com, 21 September 2011.

25 Mingjun Zhang et al., 'Nanofibers and Nanoparticles from the Insect-capturing Adhesive of the Sundew (*Drosera*) for Cell Attachment', *Journal of Nanobiotechnology*, VIII/20 (2010), pp. 1–10.

26 Nigel Hewitt-Cooper, *Carnivorous Plants: Gardening with Extraordinary Botanicals* (Portland, OR, 2016), p. 138.

6 Collecting, Growing and Conserving Carnivorous Plants

1 *Gardeners' Chronicle* (1881), quoted in Anthea Phillipps, Anthony Lamb and Ch'ien C. Lee, *Pitcher Plants of Borneo*, 2nd edn (London and Borneo, 2008), p. 25.

2 John Saul, *Catalogue of Plants for the Spring of 1884* (Washington, DC, 1884).

3 James Pitcher and W. Albert Manda, *Descriptive Catalogue of New and Rare Seeds, Plants and Bulbs* (Short Hills, NJ, 1892).

4 Wm F. Bassett & Son, *Catalogue of the Bellevue Nursery* (Hampton, NJ, 1897).

5 F. Weinberg, *Cacti, Novelties, Odd and Rare Plants* (Woodside, NY, 1906).

6 Ibid.

7 Siebrecht & Son, *General Illustrated and Descriptive Hand Book, New Rare and Beautiful Plants* (New York, 1908).

8 E. C. Robbins, Gardens of the Blue Ridge, *Special Trade Prices for July and August Acceptance* (Pineola, NC, 1924).

9 See International Union for Conservation of Nature, www.iucn.org, accessed 1 August 2017.

10 Nigel Hewitt-Cooper, *Carnivorous Plants: Gardening with Extraordinary Botanicals* (Portland, OR, 2016), p. 140.

11 Wilhelm Barthlott et al., *The Curious World of Carnivorous Plants: A Comprehensive Guide to Their Biology and Cultivation* (Portland, OR, and London, 2007), p. 149.

12 Hewitt-Cooper, *Carnivorous Plants*, p. 106.

13 Peter D'Amato, *The Savage Garden: Cultivating Carnivorous Plants*, revd edn (Berkeley, CA, 2013).

14 Phillipps et al., *Pitcher Plants of Borneo*, p. 271.

15 Carl Zimmer, 'Fatal Attraction', *National Geographic*, CCXVII/3 (2010), pp. 80–94.

16 Larry Mellichamp and Paula Gross, *Bizarre Botanicals: How to Grow String-of-Hearts, Jack-in-the-Pulpit, Panda Ginger, and Other Weird and Wonderful Plants* (Portland, OR, and London, 2010), p. 47.

17 Phillipps et al., *Pitcher Plants of Borneo*, p. 271.

18 Ibid.

19 David E. Jennings and Jason R. Rohr, 'A Review of the Conservation Threats to Carnivorous Plants', *Biological Conservation*, CXLIV (2011), p. 1357.

20 Phillipps et al., *Pitcher Plants of Borneo*, p. 272.

21 Adam T. Cross et al., 'Seed Reproductive Biology of the Rare Aquatic Carnivorous Plant *Aldrovanda vesiculosa* (Droseraceae)', *Botanical Journal of the Linnean Society*, CLXXX (2016), pp. 515–29.

22 See http://nasarracenia.org, accessed 1 September 2017.

23 S. Leonard Bastin, 'Carnivorous Plants of the Future', *Scientific American* (18 December 1909).

24 Ibid.

25 Ibid.

Further Reading

Bailey, Tim, and Stewart McPherson, *Dionaea: The Venus's Flytrap* (Poole, 2012)

Barthlott, Wilhelm, et al., *The Curious World of Carnivorous Plants: A Comprehensive Guide to Their Biology and Cultivation* (Portland, OR, and London, 2007)

D'Amato, Peter, *The Savage Garden: Cultivating Carnivorous Plants*, revd edn (Berkeley, CA, 2013)

Darwin, Charles, *Insectivorous Plants* (London, 1875)

Greyes, Natch, *Cultivating Carnivorous Plants* (North Charleston, SC, 2015)

Hewitt-Cooper, Nigel, *Carnivorous Plants: Gardening with Extraordinary Botanicals* (Portland, OR, and London, 2016)

Lloyd, Francis Ernest, *The Carnivorous Plants* (Waltham, MA, 1942)

Mabey, Richard, *The Cabaret of Plants: Botany and the Imagination* (London, 2015)

McPherson, Stewart, *Carnivorous Plants and their Habitats*, vols I and II (Poole, 2010)

Phillipps, Anthea, Anthony Lamb and Ch'ien C. Lee, *Pitcher Plants of Borneo*, 2nd edn (London and Borneo, 2008)

Rice, Barry, *Growing Carnivorous Plants* (Portland, OR, and London, 2006)

Robinson, Alastair, et al., *Drosera of the World*, vols I–III (Poole, 2017)

Associations and Websites

Societies

ASSOCIATION FRANCOPHONE DES AMATEURS DE PLANTES CARNIVORES
(FRANCE)
www.dionee.org

ASSOCIAZIONE ITALIANA PIANTE CARNIVORE (ITALY)
www.aipcnet.it

CARNIVORA (NETHERLANDS)
www.carnivora.nl

THE CARNIVOROUS PLANT SOCIETY (UK)
www.thecps.org.uk

GESELLSCHAFT FUR FLEISCHFRESSENDE PFLANZEN (GERMANY)
carnivoren.org

INSECTIVOROUS PLANT SOCIETY (JAPAN)
ips.2-d.jp

INTERNATIONAL CARNIVOROUS PLANT SOCIETY
www.carnivorousplants.org

NEW ZEALAND CARNIVOROUS PLANT SOCIETY (NEW ZEALAND)
www.nzcps.co.nz

VICTORIAN CARNIVOROUS PLANT SOCIETY (AUSTRALIA)
www.vcps.org

Conservation

ARK OF LIFE
Ex-situ conservation of Nepenthes and other carnivorous plants
www.arkoflife.net

CITES CONVENTION ON INTERNATIONAL TRADE IN ENDANGERED SPECIES OF
WILD FLORA AND FAUNA
www.cites.org

IUCN (INTERNATIONAL UNION FOR CONSERVATION OF NATURE)
www.iucn.org

IUCN CARNIVOROUS PLANT SPECIALIST GROUP
Sir David Attenborough, Patron
www.iucn-cpsg.org

NORTH AMERICAN SARRACENIA CONSERVANCY (NASC)
Conservation of North American pitcher plants
http://nasarracenia.org

Websites and Forums

BARRY RICE CARNIVOROUS PLANTS
www.sarracenia.com

THE CARNIVORE GIRL
www.thecarnivoregirl.com

CARNIVOROUS PLANT FORUM, UK
www.cpukforum.com

FLY TRAP CARE, CARNIVOROUS PLANTS
www.flytrapcare.com

NATCH GREYES CARNIVOROUS PLANTS
ngcarnivorousplants.tumblr.com

THE PITCHER PLANT PROJECT, ROB CO
www.thepitcherplantproject.com

TERRA FORUMS, CARNIVOROUS PLANT DISCUSSION
www.terraforums.com

TOM'S CARNIVORES
www.carnivorousplants.co.uk

Nurseries

CALIFORNIA CARNIVORES (USA)
www.californiacarnivores.com

COLLECTORS CORNER (AUSTRALIA)
www.collectorscorner.com.au

HAMPSHIRE CARNIVOROUS PLANTS (UK)
www.hampshire-carnivorous-plants.co.uk

SARRACENIA NORTHWEST (USA)
www.cobraplant.com

TRIFFID PARK (AUSTRALIA)
www.triffidpark.com.au

Gardens/Collections

THE HUNTINGTON BOTANICAL GARDENS (USA)
www.huntington.org

JARDIN DES PLANTES (PARIS)
www.jardinesplantes.net

KEW ROYAL BOTANIC GARDENS (LONDON)
www.kew.org

ROYAL BOTANIC GARDENS (SYDNEY)
www.rbgsyd.nsw.gov.au

SINGAPORE BOTANIC GARDENS (SINGAPORE)
www.sbg.org.sg

YUENOSHIMA TROPICAL GREENHOUSE DOME (JAPAN)
www.yumenoshima.jp/english.html

Acknowledgements

I would like to thank the Victorian Carnivorous Plant Society, Triffid Park, Collectors Corner, Richard Allen, Tim Entwisle and Randall Robinson. Also, a big thanks to my family for their constant support.

Photo Acknowledgements

The author and publishers wish to express their thanks to the below sources of illustrative material and/or permission to reproduce it. Some locations of artworks are also given below, in the interests of brevity:

From F. A. Brockhaus, *Brockhaus' Konversations-Lexikon* (Leipzig, 1892): p. 14; from J. W. Buel, *Sea and Land* (Toronto, 1887): p. 118; V. A. Pogadaev (Wikimedia commons): p. 159; © Dan Corson: p. 154; from Horace Cox, *Impecuniosus, Unasked Advice: A Series of Articles on Horses and Hunting* (London, 1872): p. 48; from *Curtis's Botanical Magazine* (London): pp. 104 (1847), 66 (1849), 108 (1858), 68 (1890), 65 (1905), 74 (1915); © Will Elder: p. 175; from John Ellis, *A Botanical Description of a New Sensitive Plant, Called Dionaea muscipula: or, Venus's Fly-trap* (London, 1770): p. 87; from R. H. Francé, Walther Ulrich Eduard Friedrich Gothan and Willy Lange, *Das Leben der Pflanze* (Stuttgart, 1906): p. 97; © Jason Gamrath/Lumina Studio: p. 153; from *The Gardeners' Chronicle and Agricultural* (London, 1872): p. 171; from Howard R. Garis, 'Professor Jonkin's Cannibal Plant', *The Argosy* (August 1905): p. 121; from *Die Gartenlaube* (Leipzig, 1875): p. 112; from Ernst Haeckel, *Kunstformen der Natur* (Leipzig and Vienna, 1904): p. 189; © Paul Hill: p. 155; © Jane Ianniello: p. 150; from *L'Illustration Horticole*, vol. 33 (Paris, 1886): p. 92; from *l'Illustration Horticole*, vol. 35 (France, 1888): p. 60; from Anton Joseph Kerner von Marilaun and Adolf Hansen, *Pflanzenleben: Erster Band: Der Bau und die Eigenschaften der Pflanzen* (Leipzig, 1913): p. 137; Kew Gardens: p. 110; © Evan Kolker: p. 152; from Hans Kraemer, *Der Mensch und die Erde* (Berlin, 1906): p. 180; from John Lindley and Joseph Paxton, *Paxton's Flower Garden*, vol. 1 (1850)/photo Wellcome Library: p. 94; from Mathias de L'Obel, *Nova Stirpium Adversaria* (London, 1576): p. 109; from George Loddiges, *The Botanical Cabinet Consisting of Coloured Delineations of Plants from all countries* (London, 1818): p. 71; from Hermann Julius Meyer, *Meyers Konversations-Lexikon*, vol. 17 (Leipzig, 1897): p. 8; from *The Ogden Standard-Examiner* (26 September 1920): p. 117; from Frank R. Paul, *Famous Fantastic Mysteries*, vol. 2 (America, 1940): p. 123; from William Robinson, *Flora and Sylva*, vol. 3, no. 32 (London, 1905): p. 48; from *Scientific American* (1909): p. 198; from Spencer St John, *Life in the Forests of the Far East; or, Travels in northern Borneo*, vol. 1 (London, 1863): p. 84; from Robert John Thornton, *The Temple of Flora* (London, 1807): p. 136;

Index

Page numbers in *italics* refer to illustrations